From *Ejido* to Metropolis, Another Path

American University Studies

Series XXI
Regional Studies

Vol. 6

PETER LANG
New York • San Francisco • Bern • Baltimore
Frankfurt am Main • Berlin • Wien • Paris

David Cymet

From *Ejido* to Metropolis, Another Path

An Evaluation on *Ejido* Property Rights and Informal Land Development in Mexico City

PETER LANG
New York • San Francisco • Bern • Baltimore
Frankfurt am Main • Berlin • Wien • Paris

Library of Congress Cataloging-in-Publication Data

Cymet, David.
From ejido to metropolis, another path : an evaluation on ejido property rights and informal land development in Mexico City / David Cymet.
p. cm. — (American university studies. Series XXI, Regional studies; vol. 6)
Includes bibliographical references .
1. Land use, Urban—Mexico—Mexico City. 2. Squatter settlements—Mexico—Mexico City. 3. City planning—Mexico—Mexico City. 4. Ejidos—Law and legislation—Mexico—Mexico City. I. Title. II. Series.
HD330.M53C96 1992 333.33'7'097253—dc20 92-10122
ISBN 0-8204-1908-7 CIP
ISSN 0895-0482

Die Deutsche Bibliothek-CIP-Einheitsaufnahme

Cymet, David:
From Ejido to Metropolis, another path : an evaluation on Ejido property rights and informal land development in Mexico City / David Cymet.—New York; Berlin; Bern; Frankfurt/M.; Paris; Wien: Lang, 1992
(American university studies : Ser. 21, Regional studies ; Vol. 6)
ISBN 0-8204-1908-7
NE: American university studies/21

The paper in this book meets the guidelines for permanence and durability of the Committee on Production Guidelines for Book Longevity of the Council on Library Resources.

Printed in the United States of America.

CONTENTS

TABLES

LIST OF FIGURES

LIST OF ACRONYMS

AURIS	Instituto de Acción Urbana e Integración Social.
CCCP	Comisión de Conurbación del Centro del País.
CNC	Confederación Nacional Campesina.
CNOP	Confederación Nacional de Organizaciones Populares.
COCODA	Comisión Coordinadora de Desarrollo Agroforestal.
COCODER	Comisión Coordinadora de Desarrollo Rural.
CODEUR	Comisión de Desarrollo Urbano.
CONAPO	Comisión Nacional de Población.
CONASUPO	Compania Nacional de Subsistencias Populares.
COPEVI	Centro Operacional de Vivienda y Poblamiento.
COPLADE	Comisión para la Planeación del Estado.
CORETT	Comisión de Regularización de la Tenencia de la Tierra.
CTM	Confederación de Trabajadores de México.
DAAC	Departamento de Asuntos Agrarios y Colonización.
DART	Departamento de Reservas Territoriales (SM).
DDF	Departamento del Distrito Federal.
DF	Distrito Federal.
DGDUV-SM	Dirección General de Desarrollo Urbano y Vivienda
DO	Diario Oficial
ECLA	Economic Commission for Latin America (UN).
SM	State of Mexico (Estado de México).
FIDEURBE	Fideicomiso de Desarrollo Urbano.
FIFONAFE	Fideicomiso Fondo Nacional de Fomento Ejidal.
FONAFE	Fondo Nacional de Fomento Ejidal.
FOVI	Fondo de la Vivienda.
FOVISSSTE	Fondo de la Vivienda de los Trabajadores al Servicio del Estado.
IMSS	Instituto Mexicano del Seguro Social.
INDECO	Instituto de Desarrollo de la Comunidad.
INEGI	Instituto Nacional de Estadistica, Geografía e Informatica.
INFONAVIT	Instituto Fondo Nacional de la Vivienda para los Trabajadores.
INPI	Instituto Nacional de Protección a la Infancia.
INV	Instituto Nacional de la Vivienda.
LFRA	Ley Federal de la Reforma Agraria.
MAMC	Metropolitan Area of Mexico City.
MRC	Movimiento Restaurador de Colonos.
PEMEX	Petroleos Mexicanos.
PNDU	Plan Nacional de Desarrollo Urbano (1978).
PND	Plan Nacional de Desarrollo (1983).
PRI	Partido Revolucionario Institucional.
PVF	Programa Financiero de la Vivienda.
SAHOP	Secretaría de Asentamientos Humanos y Obras Públicas.
RHP	Renovación Habitacional Popular

SEDUE	Secretaría de Desarrollo Urbano y Ecología.
SIAP	Sociedad Interamericana de Planificación.
SMP	Sociedad Mexicana de Planificación.
SPP	Secretaría de Programación y Presupuesto.
SRA	Secretaría de la Reforma Agraria.
ZUE	Zona Urbano Ejidal.

CHRONOLOGICAL TABLE

Presidents, DF Mayors and SM Governors Since 1940

Presidents	DF Mayors	SM Governors
1940-1946		
Manuel Avila Camacho	Javier Rojo Gomez	Wenceslao Labra Alfredo Zárate Albarrán José Luis Gutiérrez Isidro Fabela
1946-1952		
Miguel Aleman V.	Fernando Cásas Aleman	Alfredo del Mazo
1952-1958		
Adolfo Ruíz Cortines	Ernesto P. Uruchurtu	Salvador Sánchez Colín
1958-1964		
Adolfo López Mateos	Ernesto P. Uruchurtu	Gustavo Baz P.
1964-1970		
Gustavo Díaz Ordaz	Ernesto P. Uruchurtu Alfonso Corona del Rosal	Juan Fernández Albarrán
1970-1976		
Luís Echeverria A.	A. Martinez Dominguez Rodolfo Sentíes	Carlos Hank González
1976-1982		
José Lopez Portillo	Carlos Hank González	Jorge Jiménez Cantú
1982-1988		
Miguél de la Madrid	Ramon Aguirre V.	Alfredo del Mazo Alfredo Baranda García
1988-1994		
Carlos Salinas de Gortari	Manuel Camacho Solís	Mario Ramón Beteta Ignacio Pichardo Pagaza

From *Ejido* to Metropolis, Another Path

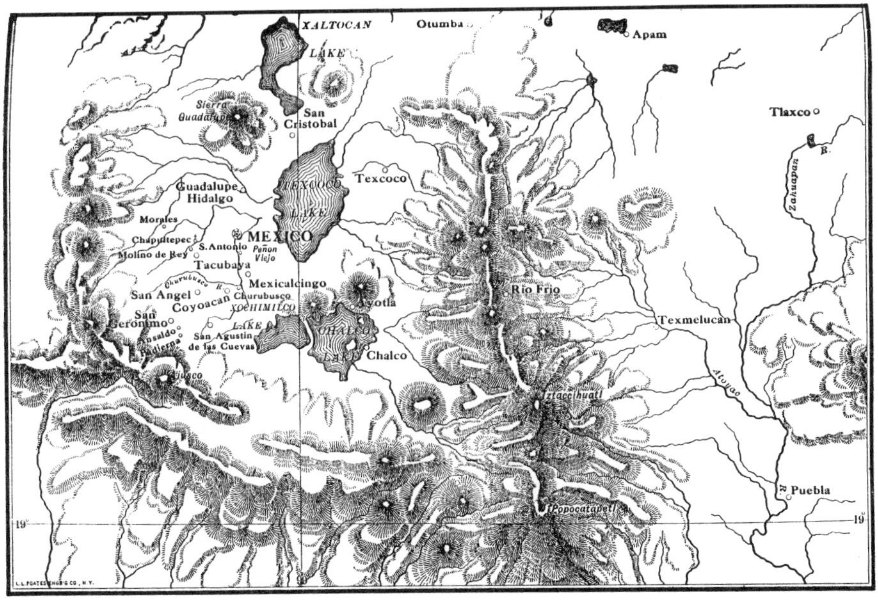

THE VALLEY OF MEXICO

Chapter I

THE FAILURE OF PLANNING IN MEXICO CITY

THE GROWTH OF MEXICO CITY

On a global scale, Mexico City is probably the most representative example of a primate city in a developing country facing a process of population growth unprecedented in magnitude and velocity. Within the last several decades, it is becoming the most populous city in the world, with its metropolitan population rapidly moving towards the twenty million mark. This is the combined result of natural population growth and the redistribution of population in the country, which has brought large masses of migrants to the city from the countryside. As a result, a heavy strain has been placed on the city's finances and natural environment, especially land and water.

This growing concentration of population is correlated to an even greater concentration of capital and economic activities in the metropolitan area of Mexico City (MAMC) which is reflected in almost every economic indicator. Paradoxically, there is also a concentration of poor people in the metropolitan area. In a manner typical of many developing countries, a transfer of rural poverty to the metropolitan area has taken place. This transfer has become more visible over time, as the proportion of the poor in the city continues to rise. Many of the poor are migrants fleeing poverty in the rural areas of central Mexico and seeking a better chance of survival in the capital.

While there is debate about the effects of such a momentous process of demographic and economic concentration on such social phenomena as income distribution, unemployment, family stability, political stability, and crime, there is little doubt concerning its impact on the physical environment. An accelerated concentration of population of such magnitude in an existing city, even without the large numbers of poor migrants being its main component, is bound to produce very heavy externalities of the kind already well known in the metropolises of the developed world. One could expect such consequences as environmental contamination, traffic congestion, and physical overcrowding, even under the best of circumstances. These externalities are being exacerbated by Third World scarcity, while additional problems more typical of the Third World, such as the irregular settlements, have likewise reached critical limits.

The complex process of national population growth and redistribution is being compounded internally by a powerful movement of population redistribution within the MAMC. The metropolitan region is undergoing locally a major internal process of population redistribution from the crowded and more costly inner city to the farther out expanding periphery. This trend which became visible in the sixties was greatly strengthened after the 1985 earthquake, which displaced many people in the inner city. In the 1980-90 decade alone more than an half million people left the inner boroughs of the DF for the outer periphery adding extra population pressure on the urban fringes.

The expansion on the urban periphery is being increasingly dominated by unplanned irregular development. The population growth rate of the irregular areas in the urban periphery is estimated to be two to three times greater than the city growth rate as a whole. This process of informal urbanization is characterized by its lack of official approval and property titles, by makeshift housing, and, what is most important, by the absence of local utilities and human services that are necessary for an urban community to function. Some early studies estimated the proportion of the population living on this type of settlement to have increased from 14% in the early forties to over 50% in the eighties, however, more reliable recent assessments estimate that proportion to be around 65% of the total population of the MAMC.[1]

Development of the irregular settlements in the urban periphery is mostly taking place on the land of the *ejidos*, the peasant communities surrounding Mexico City whose historical roots date back to pre-Colombian times. The Mexican revolution (1911-20) gave new life to this ancient institution as a means to carry out land reform in the country. Many *ejido* communities were created and endowed with land taken from the large estates, the *haciendas*. The property rights, however, which were granted to the *ejido* communities over their land endowments were greatly restricted in the Mexican constitution of 1917: these could not be marketed nor mortgaged.

How and why irregular land development is dominating the process of urban expansion are important questions which have been raised in the literature. The objective of this book is to challenge some of the assumptions on the origins of the irregular settlements widely held in the literature and establish the argument that the greatly diminished property rights of the *ejidos,* the peasant communities surrounding Mexico City, played a major role within the context of the national economic policies in creating this situation. As a result, this discussion should contribute to recognize the crucial importance of secure property rights in economic development in general and urban development in particular.

THE HISTORY OF URBAN PLANNING IN MEXICO CITY

Most of the city's expansion and particularly that of its irregular settlements has occurred through indiscriminate occupation of land. Most often this expansion is accomplished with little regard for the land's fitness for urban use, its degree of risk to human life and property, or its ecological role. This expansion has resulted in the destruction of environmentally sensitive areas and in a great increase in the city's vulnerability to natural and technological disasters. The occupation of the highly vulnerable, inhospitable, dry bed of Lake Texcoco by one and a half million people and the contamination of the Ajusco aquifer recharge area by irregular settlements are just two examples of this destruction.

Chaotic expansion has continued in spite of the government's long declared policy of planned urban growth. Although this policy has been expanded over time to assume greater areas of responsibility, it has repeatedly failed to achieve its objectives. In earlier times these policy commitments were exclusively limited to physical planning. Their aim was to guide some aspects of spatial growth. In more recent times the policies were expanded to include natural population growth and population redistribution in an effort to attack the roots of the city's growth problems, not only the effects.

As early as 1936, just before the city's growth boom began—when it still was completely contained within the jurisdiction of the Distrito Federal (DF)—the government established a modern policy commited to urban planning. A Planning and Zoning Law in Mexico City was enacted as the result of the pioneering effort of Carlos Contreras. A special planning office was created within the city government, and a master plan for the city containing an industrial zoning plan and a transportation plan was officially adopted.

The city government in the early forties also approved advanced building and land development regulations. Higher standards were aimed at preventing developments without services, slum formation, and jerry-building. The land development regulations required, among other things, that potential developers present registered title to the property and that they set aside a monetary guarantee to make sure that all the public services be carried out. The sale of lots before the installation of all services was forbidden. The government of the DF intended to influence through these means the physical character of the city's growth by maintaining minimal standards in land development and building activity and by approving the location of the most critical urban land use.

The decade of the forties ushered in a period of growth and expansion of the city, despite the great technical difficulties and escalating costs of expanding the water-supply and sewer systems. By the end of the forties the expanding city spilled onto its northern frontier to the neighboring municipalities of the State of Mexico. A moratorium on further urban growth in the DF was issued in 1953 by a strong-handed new mayor, Ernesto P. Uruchurtu (1953-66). His ban did not stop growth but redirected it in part to the neighboring State of Mexico. Here the results were even less satisfactory because of the far greater limitations in the fiscal and administrative resources of the state government. Within the DF itself, much irregular settlement activity took place on land of the local *ejidos*, or peasant communities, despite the ban. More stringent and advanced land development regulations were also adopted by 1958 in the State of Mexico. Like those of the DF, they were intended to bring about well-serviced industrial and residential subdivisions.

In the seventies there were important new developments in the government's commitment to planning. An awareness that policies of that time were leading to unprecedented population growth and rapid urbanization, unbalanced population distribution, regional, social, and economic inequalities, and centralization gave rise to a variety of programs to counter these effects. To check population growth, a national population planning policy was formulated and implemented by the Mexican government during the Echeverria administration (1970-76). This policy led to the adoption of a national population law in 1973 and the creation of the National Population Council, CONAPO, whose main goal has been the reduction of the population growth rate through voluntary means. Although that program was national in scope, it implicitly aimed at the reduction of the capital's overall growth rates through a decrease in natural growth and by reducing its external migratory growth source.

To counter the urbanization and centralization trend, the government revived some long-existing programs in the agricultural sector. It also introduced some new programs in the seventies aimed at reducing rural exodus by creating incentives for the peasants to stay in the rural areas. In a similar fashion, the governmental industrial sector justified many of its development projects, such as the building of industrial ports and industrial parks, on grounds that these would reroute migration of people and capital away from Mexico City and the two other major cities, Guadalajara and Monterrey. An initiative to move certain government agencies out of the capital remained inoperative.

The preparations for the incoming presidential administration of José López Portillo (1976-82) at the end of the Echeverria presidency were marked by a variety of momentous initiatives. Important commitments were made at the federal level for planned urban development. A constitutional amendment to article 27 of the 1917 Constitution that explicitly made the planned development of the population centers a national responsibility was approved on the fiftieth anniversary of the Querétaro Constitution.[2]

The constitutional amendment became the basis for the promulgation of the Human Settlements Law (Ley General de Asentamientos Humanos) adopted in the same year. The Human Settlements Law defined further in a broad way the instruments that were to be used to carry out the urban planning activity mandated in the constitutional amendment. This instrumental role was assigned to urban development plans at the national, state, and municipal level. They were to serve as the basis of legally binding declarations called *"Declaratorias de Usos, Destinos y Reservas"* defining for each city its public and private land uses and its land reserves for future urban growth. For conurbated areas extending over more than one political jurisdiction, as in the case of Mexico City, special provisions requiring the establishment of inter-jurisdictional planning commissions and plans were adopted.

That same year a new national ministry for regional and urban planning, the Ministry for Human Settlements and Public Works (Secretaría de Asentamientos Humanos y Obras Publicas, SAHOP, later renamed SEDUE) was created, putting planning and public works under the same authority. Significant financial and human resources were made available. The responsibility for creating urban land reserves was transferred to the new Ministry of Human Settlements to allow the government to provide enough land for its public housing projects and influence the character of land development. An urban development plan was produced by SAHOP at the national level (PNDU). This plan envisioned a decentralized system of cities in the country to counter the centralization trend toward Mexico city.

The National Urban Development Plan (PNDU) intended to create a hierarchical system in which the middle sized cities would play a much more important economic and demographic role than before. The PNDU defined three types of different urban growth policies to be applied to the cities of the urban system depending on their potential: control, consolidation, or growth stimulation. For many of the middle-sized and small cities possessing adequate resources, it recommended growth stimulation. For those cities facing natural resources limitations, it recommended a policy of consolidation. For Mexico City, Guadalajara, and Monterrey, the three major cities, it recommended a control policy aimed at limiting growth. In the case of Mexico City this recommendation included the total ban on new industries in this area, the creation of new growth poles in the coastal areas (industrial ports), and a population redistribution policy aimed at diverting migration away from the capital to other destinations.

A regional interstate planning commission was soon created for the Mexico City conurbation under the name of Comisión de Conurbación del Centro del País (CCCP) in compliance with the Human Settlements Law. In 1978 the Commission produced a regional plan which articulated a scheme to control Mexico City's growth through decentralization of many of the capital's activities. Capital cities of the surrounding neighboring states as well as new towns to be established in the region were to assume this redistributed burden. The building of a freeway and railroad ring to link these cities and the creation of concentric belts of built-up areas, separated by open space, were planned to facilitate the decentralization movement.

Meanwhile, the DF government reorganized its planning apparatus and made the existence of a master plan and a zoning system legally mandatory. A new master plan that defined three descending levels of planning—the DF, the sixteen *delegaciones* or boroughs, and the neighborhoods—was issued. Using general resource and ecologic aptitude criteria for the first time, the Plan Director defined three zoning areas in the DF: first, an urban core to be given greater density and restructured in nine urban centers and several high density corridors; second, a buffer area in the southern urban periphery to serve as a growth containment belt that would be subject to decreasing development to maintain its rural character by means of restricting land use to intensive agriculture, recreational areas, and parks; and third, an ecological reserve area covering the rest of the Federal District, containing villages, farms, national parks and forests to be kept as a natural conservation and reforestation area by means of strict controls.

The neighboring State of Mexico, into which the city had expanded, also produced an urban development plan. While it basically followed the guidelines of the regional plan of the Conurbation Commission, it was ambivalent on the limitation of growth in the immediate areas surrounding the Federal District and the removal of the incentives to industrial growth.

The eighties witnessed some important changes in planning policy. The new administration discarded the separate plans for the various government sectors of the preceding government, including the National Urban Development

Plan of SAHOP (PNDU). The sectorial plans were turned into subsidiary programs of one national development plan, the Plan Nacional de Desarrollo (PND). The PND was placed in the hands of SPP, the Ministry of Budget and Programming, which had gained greater political ascendancy over the other government ministries.

In what constitutes a most important policy reversal, the regional decentralization policies of the former administration were discarded by the new government and the creation of new development poles such as the industrial ports was suspended. Control over public works was transfered from SAHOP/SEDUE to an independent ministry. The strategic importance of the metropolitan area for economic growth was reaffirmed. The PND abandoned the widely-accepted definition of decentralization as the geographic redistribution of population and activities from the capital and reinterpreted it in a political sense as a redistribution of power from the central to the local governments. The term deconcentration was introduced and local deconcentration of the MAMC to its immediate surrounding metropolitan area was adopted instead of decentralization. The PND suggested that investment be channeled to the peripheral areas of the city, especially to the neighboring municipalities of the State of Mexico, and recommended extension of the Metro to them. It also recommended stricter land use controls, increased urban densities, and greater community participation in the planning process through neighborhood organizations.

A new development program for the central region and the metropolitan area of Mexico City was jointly formulated in 1983 by the office of the President, the DF government, and the five neighboring state governments. Considering the limited availability of water and the critical environmental conditions of the metropolitan region a selective industrial development policy was suggested by the program, favoring non-contaminating industries with low water consumption ratios. The program recommended that limitations be placed on the horizontal expansion of the urban area to avoid further environmental disruption. Urban densification and filling-in of vacant spaces to maximize use of existing infrastructure were recommended as a strategy to absorb population growth and prevent expansion of the urban area. A growth containment policy was formulated for the southern rim of the DF by means of a rural buffer zone which was officially incorporated into the 1987 Master Plan. The creation of a special organization (COCODA, 1986, renamed COCODER in 1989), the adoption of a zoning code for the buffer zone and the acquisition of land by the government were seen as the means to implement the buffer concept.

THE FAILURE OF URBAN PLANNING IN MEXICO

Although one can point to a policy response on the part of the Mexican government to confront the metropolitan growth process of the capital city, particularly in the last four presidential administrations (1970-92), there can be little doubt that it has been largely unsuccessful in achieving its declared goals. From the standpoint of population growth and population redistribution, no reduction in the city's natural and migratory growth, nor any visible decentralization of population or economic activities, has taken place. A marked deterioration in the environmental conditions of the region has set in which is considered critical by world standards. The city is rapidly advancing on its rural periphery, including the so called buffer zone at the southern rim of the city. Time and again the containment policies failed to achieve their objectives. By 1980 five more municipalities of the SM were taken over by unplanned urbanization and by 1990 another 11 municipalities of the SM joined the MAMC conurbation, making a total of 27 municipalities within that neighboring state. Chaotic urban expansion continues in spite of the government's declared policies in favor of planned urban growth.

The failure of urban planning in Mexico City has frustrated policy makers and analysts. It has become a subject of academic inquiry, eliciting some responses by Third World analysts including Peter M. Ward, Tim Campbell and others.[3] The explanation of this failure is seen as an important question which has a great significance from an analytic and policy standpoint, not only for crisis-ridden Mexico City, but also for Third World cities experiencing rapid growth as well.

These authors have put forward various explanations for the failure of the planning system, particularly after 1976 when it received great backing in the form of a Ministry for Urban Planning. Their diagnosis can be summarized as follows:

— In a society of closed-door decisions, these are reached by the top politicians on the basis of particularist criteria protecting group interests. For politicians, political control and patronage is important, not the efficiency of development. Politicians need to be able to exercise discretionary control over the allocation of resources and to be able to act on the basis of their personal decisions to accommodate competing economic and political alliances. Planning poses a threat to traditional forms of political mediation that are based on patronage. They therefore do not allow planning to become institutional and continue to keep it personal.

— Physical planning is used as an ideological tool to enhance state control, rather than as a regulatory framework for urban development. "Approved" plans can serve to justify political actions: to secure loans from international agencies, to justify or increase budgetary allocations from the cabinet or to legitimize decisions already made. Plans and regulations have been waived or ignored when confronted by the political or economic imperatives of the day. The centralized character of the political system makes planning more of a record keeping exercise.[4]

— Lack of public participation as a result of personal decision making by top politicians has weakened urban plans. Public participation has been merely nominal and is used only as a mechanism for exerting greater control. The last thing that decision makers want is informed criticism. Planning is feared for its greater dissemination of information. Ward, in particular, sees lack of public participation in the planning process as the most important single explanation of the failure of the planning system in Mexico. "If planning" says Ward, "as an open, democratic and technical process is ever to exist in Mexico, then the key is almost certainly one of public participation."

— The action among Mexican urban planning agencies has been piecemeal, dissonant, even at cross-purposes, making it difficult for any of them to exercise power. The many offices with responsibility for planning whose fortune and existence rises and wanes with every new government, compete with each other for their duties and power, and are played off against each other by the higher authorities. Competition between the bureaucratic political factions, the *camarillas,* is the hallmark of the Mexican government system. Even during the most active phase of planning (1976-1982), the PNDU foundered because the cooperation of the other government bureaus was not forthcoming. The government has not sought to create a city wide authority. Cooperation between planning agencies on either side of the Federal District boundary is essential but very unlikely, as they normally belong to different *camarillas.* The *Plan Director* has not addressed the problem of complementarity with planning initiatives on the other side of the DF boundary line.

— Plans lack continuity from one presidential administration to the next.

— There is an absence of specific and concrete planning objectives. Evaluation of regional development policies, plans, and programs becomes impossible when it only addresses vague, broad strategies and general concerns instead of concrete objectives.[5]

— Plans and planning are still emerging from their status as formalized images of a future, rather than as rigorous tools to facilitate improvements or secure objectives. The plans lack elements crucial for policy implementation. Much of the emphasis is on goal-setting rather than process building. They prescribe what to do, instead of presenting guidelines on how to achieve development goals, or how to introduce socio-economic and environmental changes and impact-assessment procedures in policy design and decision making. The plans are indicative, they do not establish a process to evaluate alternatives or to assess the feasibility of proposals.

— The plans lack elements crucial for policy implementation. They do not contain growth controls as they are known in the United States. Controls based on land use criteria were never implemented, due to the lack of enforcement capabilities, especially in the critical areas such as the buffer and conservation zones. Negative controls of land use, taxation, and zoning, to restrict location in undesirable sites, are absent. Political authorities have much harder bullets to bite, beyond positive incentives—such as stiffer fines, more rigorous regulatory mechanisms, and wholesale revisions in land use laws—before effective controls, much less decentralization can be realized.

— Planning in the DF, and more so in the State of Mexico is poorly funded. Municipal governments are still far from being able to resist the corrosive influences of bought favors, which largely demolish urban plans, land use controls, and industrial location regulations.

— The promise and technical potential of the Mexican planning system is not matched by its political and social readiness to change. The political power concentrated in the center largely prevents any serious attempt to regulate industrial location. Withdrawal of industrial investment from the MAMC to other locations as recommended by the PNDU never took place.

AN EVALUATION OF THE LITERATURE: TWO MAIN PROBLEMS

Although there is no reason to disagree specifically with any of the formulations given by these authors to explain the failure of the urban planning policies in Mexico City, taken as a whole these formulations leave out crucial explanatory elements. First, there is a need to view this failure in a more comprehensive general way. Said failure should be viewed within the general framework of the Mexican economic and political model. Only then does it become possible to understand the subordinate role that the urban planning policies played with respect to the national economic and political policies. Such an analysis will reveal that the planning policies could hardly have been effective given their chronic inconsisteny with the economic and political policies.

Second, these authors have failed to perceive in their diagnostics the major role that the property rights of the *ejidos*—the peasant communities surrounding Mexico City—have played within the framework of the national economic and political policies as a major institutional obstacle to physical planning in the urban peripheries. The thesis is put forward that the *ejido's* present institutional arrangements, on which most of the irregular settlements have developed, have acted as a powerful barrier that has prevented the urban planning system from fulfilling its function in guiding the city's expansion process. The facts summoned lend support to the contention that the highly restricted and diminished land property rights of the surrounding *ejido* communities operating within the context of the national economic policies pursued for over fifty years, had a decisive influence on the character of urban land development in the periphery of Mexican cities. They are a crucial link to explain the extraordinary proliferation of the irregular settlements.

Resolving the *ejido* dilemma in the periphery of rapidly growing population centers, such as in the specific case of Mexico City, is a necessary, although by no means a sufficient, condition to create the opportunity for urban planning to fulfill its function in Mexico. One must deal in such a situation with the total environment of planning, of which the *ejido* is only one element, albeit a very important one. Much as it is true that many other problems have to be solved before urban planning is allowed to be effective, not least those previously pointed out by these authors or those originating in the political economic model. It is also true that unless the obstacle of the present *ejido* arrangements in the growth area of the city is removed, no policy to implement urban planning can aspire to be successful.

The ensuing discussion will now begin by examining the national economic policies as they relate to Mexico City growth and then proceed to visualize the role that the undefined *ejido* property rights played in determining the character of the city's expansion.

THE HISTORY OF THE *EJIDO*

Many Mexican cities, but in particular the fast growing capital, face a peculiar problematic situation unique to that country: they encounter in their expansion on their rural periphery a static, highly restricted form of communal land property belonging to agricultural communities known as *ejidos*. This communal land is at present the main source of rural land for the city's expansion; however, the restrictions attached to the property rights of this land conflict with the cities' growth needs. Since private or government-owned land, constitute a decreasing fraction of the total stock of land available in the rural areas of the Distrito Federal and the neighboring municipalities of the State of Mexico, the importance of the *ejidos* and the legal and institutonal character of the *ejido* system will tend to increase.

This type of land is fundamentally distinguished from private property by being legally non-marketable and non-transferable. Specific legislation, whose nominal purpose is to protect *ejido* communities from losing their land under any circumstance, governs *ejido* property. Ostensibly *ejido* ownership can only be removed by presidential decree and for recognized higher public purposes. Compensation in such circumstances should be paid to the *ejido* communities affected.

While the restrictions placed on *ejido* property were originally conceived for the purpose of protecting the peasant communities from losing their land, it seemed as if these same restrictions were also ideally suited for an ordered process of planned urban growth, as land for urban development could be kept out of circulation, and made available at the proper time and place by government expropriation and compensation, according to each city's growth needs. Others were convinced of its potential for implementing official programs to supply land for housing the urban poor.[6]

However, when one looks at the actual results in the periphery of the most representative of these population centers, Mexico City, the very opposite is true. The *ejidos* in the urban peripheries have been rapidly losing their land without any tangible benefit to them, and instead of facilitating planned urban growth the *ejidos* present formidable obstacles to its implementation. *Ejido* land on the peripheries of cities in Mexico is being converted to urban uses in a chaotic way. Very similarly, *ejido* land has presented great difficulties to its use in official programs to house the low income population and has instead become the dominant site of the irregular settlements.[7]

The urbanization of *ejido* land in the urban peripheries is occurring in great measure outside of the legal order. Great parts of this land are being urbanized by devious illegal means and clandestinely converted to urban use. As a result, two largely separate land markets have been created: one formal, the other informal, with the informal market supplied by the *ejido*. The illegal or secondary land market has become a crucial factor in determining where and when the city's expansion takes place, largely blocking the planning system or normal markets from fulfilling their function.

A major program to regularize the tenure situation of the illegally alienated lands on which hundreds of irregular settlements have been established has been instituted since 1973 by CORETT, a special intergovernmental commission. CORETT has legally privatized some of these developments, but has not contributed in the least to prevent illegal settlement activity and unplanned development. On the contrary, the regularization programs have actually exacerbated the problem.

The urban conversion process of the *ejidos* as a whole, both legally and illegally, has largely turned against the *ejidatarios* themselves. It leaves them generally impoverished, without land or money. The meager funds that the *ejidatarios* can extract from the alienation of the land rarely offers them any other alternative but to become poor urban day laborers or petty merchants. The record of legal expropriation is in this respect much worse. While the land is taken away from them for major private and public investments, very often of the highest economic significance, they receive minimal compensation.

Conversion of the *ejidos* to urban use is also linked very intimately to a series of other important negative social phenomena. These include the corruption of the agrarian administrative apparatus, the confrontation between *ejidatarios* and settlers and other interests in the urban periphery, and the criminalization of the socially useful activity of land development entrepreneurship.

THE FOCUS OF THIS BOOK

We will look at the question of the *ejidos* in the urban periphery in order to shed light on how urban growth on this type of land is largely occurring outside of the framework of agrarian legality and why the *ejido* has become an obstacle to planning. It is intended to use this understanding to suggest a policy aimed at reducing various negative effects on the urban planning process.

The existing literature has established the dominant role of illegally converted *ejido* land in Mexico City's growth. It has thrown some light on the contorted means used for its illegal conversion to take place. This literature, however, has not addressed the central issue, which is really the most significant from a policy standpoint: why this illegality has become the pervasive characteristic of *ejido* land conversion in the urban periphery, in spite of all the safeguards against its occurrence and the risks and losses involved. This question is addressed here, and an answer is offered that explains this phenomenon on grounds of the *ejido's* economic self interest and moral claims.

This understanding is further applied to the wider policy issue, reminiscent of the U.S. prohibition era discussion, of whether it makes sense to maintain an unenforceable largely fictitious body of Agrarian Reform Law in urban expansion areas. An inquiry is undertaken on the possibilities of narrowing the gap between law and reality. The question is asked whether the *ejido* can play a more constructive role as a resource in a growth-responsive land and resources management system, instead of acting as a major obstacle. After discussing current thinking on this matter as reflected in the literature, the Tlaxcala SIAP-SMP meeting (1981), and in the amendments to the Agrarian Reform Law (1984), a policy proposal is formulated which aims at achieving these goals.

The policy proposal argues that as a full-fledged partner sharing risks and benefits, the *ejido* can be made to play such a role voluntarily, but, as a systematic loser, it will tend to oppose the system in spite of the tighter controls and greater penalties that are advocated in some circles and implied in the literature. The study challenges the assumption—which is really a myth unsubstantiated by law and history—that the *ejido* is a form of usufruct whereby the state is the real owner of the *ejido* land which it allows the *ejidatarios* to use. This erroneous assumption has only served to confuse the issue of *ejido* rights and conversion illegality. It is being used as an ideological weapon to undermine the *ejido* rights when it suits the interests of those who covet *ejido* land and to place the blame on the victims for the current illegality.

---

1 (Turner 1971; Ward, 1981; COPEVI, 1977; Schteingart, 1988).
2 (Diario Oficial, February 6, 1976).
3 (Ward; 1986:p. 219-235; Campbell and Wilk, 1986:p. 287-313).
4 (Ward, 1981: p. 60-61).
5 (Miller, 1988: p. 9)
6 (Angel *et al.*, 1983).
7 (Varley, 1985:p. 1, 13).

Chapter II

THE EXPANSION PROCESS OF MEXICO CITY ON ITS PERIPHERY

To be able to understand in a more comprehensive way the failure of the physical planning system to guide Mexico City's growth process, one must view the planning activity within the general framework of the policies guiding the Mexican economic and political system, and to relate the growth process to the strong social and economic forces that stand behind it.

Urban growth is a complex process that manifests itself in various dimensions, demographic, economic, and spatial, that are interactively related. This chapter intends to look at the interaction of these dimensions in the MAMC. This frame of reference is necessary to reach a clearer understanding of the role played by the *ejidos* and their property rights in the expansion process of Mexican cities which is discussed in chapter III.

MAMC GROWTH: ITS DEMOGRAPHIC DIMENSION

The extent and speed of growth of the MAMC has no parallel in the developed western world from a demographic standpoint. Even among Third World developing countries the growth is without precedent. For centuries population growth rarely went beyond one percent a year, but in the 1940's Mexico's national population net growth rate suddenly leaped to almost three percent, and has been higher in subsequent decades (table 1). This veritable demographic revolution can be traced to such factors as the discovery of antibiotics, the establishment of a national health care system, and increasing sanitary infrastructure. These factors brought about a dramatic reduction in the average mortality rate of the country as a whole, at a time when the birth indices stayed the same or even increased.

Paralleling the demographic explosion, a major population redistribution process took place. An increasing proportion of the total population settled in the limited geographical area of the capital's metropolitan region. The

MAMC's share of the country's total population rose from 2.5% in 1900 to around 20% in 1980 (table 2).

The combined action of natural increase and population redistribution produced a net annual population growth rate for the MAMC which has frequently exceeded five percent per year between 1940 and 1980. The city population doubled in periods of a little more than 10 years. Thus, while the country's population increased 3.5 times from 19,654,000 to 69,393,000 people, the population of the MAMC increased almost nine times, from 1,553,000 to over 13,726,000 people during those forty years.

Mexico City, which was not on the list of the 25 most populated cities of the world in 1940, had surpassed most of the great world cities in population by 1980, rising to third place on the list. By 1990, Mexico City was competing with the Tokio-Yokohama conurbation for the first place.

However, the data from the 1990 census show that some significant changes are occurring within this general trend of population growth. The net annual growth rate of the MAMC population registered a major fall from 4.6% in the previous decade (1970-80) to 1.0% in the last decade (1980-90). The local/national population ratio may have already reached its peak in 1980 with a value of 19.8% and seems to begin its descent (see table 2). Nevertheless, in absolute numbers, the MAMC as a whole, still shows an important net population increase of 1,322,000 people over the ten-year period from 1980 to 1990.

The overwhelming numerical dominance of the capital city with respect to the other two major cities, Guadalajara and Monterrey, is continuing to subside. The primacy index which reached its peak in 1950 with a value of 7.2 continued its downward movement to around 3.0 in 1990, indicating the growing relative importance of Guadalajara and Monterrey (table 2).

Reductions in the national natural population growth rates and changes in the migratory movements to and from the MAMC are producing this slowdown in population growth. For the first time the number of migrants coming in the MAMC is somewhat smaller than the number of those leaving. A net negative balance of (-) 223,000 migrants during the 1985-90 period is the result of 1,307,000 people leaving the MAMC against 1,085,000 people coming in.[1]

As a result of the important changes that took place in the 1980-90 decade there has been a need of revising the previous MAMC population projections that were made on a different set of assumptions. The previous UN projection of 25.6 million people for the MAMC for the year 2000 is being currently revised. The recent INEGI-CONAPO projection of 103,999,000 people for the country as a whole, would point to a probable population of around 18 million people for the MAMC at the end of the decade.

Table 1. Mexico City and National Demographic Indices

year	Birth Rate/1000 national	Birth Rate/1000 local	Death Rate/1000 national	Death Rate/1000 local	Net Growth Rate national	Net Growth Rate local
1895-00	33.7	45.0	33.9	28.0	n.a	3.2
1900-10	33.5	45.0	32.9	28.0	1.1	6.2
1910-21	31.4	45.0	25.1	28.0	- 0.5	0.4
1921-30	34.4	n.a.	25.5	n.a.	1.7	4.7
1930-40	43.5	n.a.	24.1	n.a.	1.8	4.1
1940-50	44.9	n.a.	19.4	23.0	2.7	7.2
1950-60	46.2	44.7	13.3	12.9	3.1	5.3
1960-70	43.1	41.3	8.6	n.a.	3.3	5.3
1970-80	34.0	37.0	n.a.	7.3	3.4	4.6

Sources: Senior, Clarence; Colegio de Mexico; Scott, Ian; Partida V.

Table 2. Population Growth in the Metropolitan Area of Mexico City, MAMC (1921-1980)

year	population MAMC	% national population	growth rate	international rank	primacy index [2]
1900	344,000	2.5	3.2	>25th	4.4
1910	629,000	4.1	6.2	>25th	3.9
1921	659,000	4.6	0.4	>25th	4.3
1930	1,043,000	6.3	4.7	>25th	5.7
1940	1,553,000	7.9	4.1	>25th	6.5
1950	3,124,000	12.1	7.2	17th	7.2
1960	5,236,000	15.0	5.3	15th	6.1
1970	8,765,000	18.2	5.3	7th	5.9
1980	13,726,000	19.8	4.6	3rd	3.7
1990	15,048,000	18.6	1.0	n.a.	3.0

Sources: VI-XI Censos de Población y Vivienda. Unikel, (1971); Scott, (1982).

MAMC GROWTH: ITS ECONOMIC DIMENSION

The "Push" and "Pull" Forces

What are the forces that have brought about such unparalleled urban growth?

There is a debate in the literature between two competing hypotheses, both of which find their application in the case of the MAMC: one stresses the rural push factor and the other the urban pull factor (Williamson, 1986). According to the push hypothesis of Malthusian inspiration—generally preferred by demographers—the high rate of population increase in the rural areas pressing on limited land resources while mechanization of commercial agriculture is reducing agricultural employment opportunities, is bringing about the migration to the cities, where rural migrants swell the ranks of the urban poor.

The alternative pull hypothesis—which has become the favorite of economists in more recent years—maintains instead that forces operating within metropolitan areas draw in external population. The intensity of attraction is correlated with the degree of economic development in the metropolitan areas. Scale economies and agglomeration advantages along with the degree of "urban bias," a term used to refer to the domestic policies that distort prices in favor of the urban areas, influence the magnitude of the pull forces in specific cities and the concentration of investments, which in turn attract population. In the specific case of Mexico City, both forces were operating and were accelerated by political choices.

The growth and centralization of Mexico City has been the combined result of "push" forces driving people off the land, especially in the central and southern regions of the country, where increasing population pressures on limited resources of land and water are most acute, and "pull" forces attracting people to the capital, where employment opportunities, wages, standards of living, levels of investment, and subsidies are highest. The powerful impact of these dual forces on Mexico City's growth were supplemented by the Mexican economic development programs adopted by the Mexican post-revolutionary regimes.

The Economic Development Model

Although migration to Mexico City is national in character, with representatives from all the states and cities of the country, the migratory movement originates overwhelmingly in the rural areas of the Center/South region of the country, particularly from the states of Hidalgo, Puebla, Tlaxcala, Oaxaca, Guanajuato, Guerrero, Michoacan and Veracruz. There the push forces are strongest. The lack of opportunities to earn a living compel people to leave the villages and *ejidos* of these states. According to recent estimates, 70% of the total number of migrants to Mexico City, which captures over 50% of the total migratory movement in the country, came from rural areas in these states.

It is argued in the literature that this powerful push force was built-in in the Southern and Central region in the 1920's by token agrarian reform, which allotted minimal plots of two or three hectares, or even less, to the peasants demanding land reform. These small subsistence parcels dedicated to the cultivation of maíz, with little water and less credit, could hardly have provided a decent living to the peasant families in the *ejidos*.[2]

Mounting pressure was being exerted by a rapidly growing rural population on the small subsistence plots, leading to increased migration—seasonal and permanent—to Mexico City and the United States border.[3] In spite of the decline of the rural population as a percentage of the total population, its growth in absolute numbers in a typical Third World manner continued feeding the migratory currents from these regions. In other agricultural regions of the country such as in the North, operating under a different land tenure situation where commercial agriculture for export prevails, the push forces were strongly influenced by the fluctuations in demand and prices of the international markets.

The typical answers cited in the literature[4] when the poor migrants are queried in their shacks along Mexico City's periphery on why they decided to migrate to Mexico City, refer generally to their former hopeless economic situation of unemployment, dire poverty, and even hunger. The following are typical reasons given by them for migrating: "There were no jobs," "There was not enough land to feed my family," "Everything is dead down there, so I came up here to see if I could find some work," "I had no food for my children."

Larrisa Adler Lomnitz comments: "Not even the oldest grandmothers showed sentimental yearnings for the native soil, their dominant memory seemed to be: we were ever starving, no one thinks of going back to the village. Life is easier here in the shantytown." Only 9.8% of the Lomnitz sample stated non-economic reasons as the main motive for migration. Thus, it is often argued quite convincingly that the agrarian reform instead of retaining the rural population perpetuated poverty in the central rural region, and encouraged migration to the capital.

The redistribution of population that took place as the combined result of the "push" and "pull" forces responsible for Mexico City's momentous growth brought about a concentration of poverty in the metropolitan area, with the addition of many poor migrants to the native urban poor. In a manner very typical of many developing countries, rural poverty was transferred to the metropolitan area, where it has become more visible over time, as the proportion of poor in the city continues to rise. Many of the poor migrants leaving the rural areas of central Mexico were attracted to Mexico City in spite of the fact that the industrial sector was not strong enough to supply employment for all of them.[5]

The agrarian reform not only contributed impoverished peasants to Mexico City, but also brought about the withdrawal of investment capital during the twenties from the agricultural estates threatened by agrarian reform to the city in search of greater security and higher profits. Although the agrarian reform had not yet completely destroyed the *hacienda* system, the system was damaged badly enough during the second half of the twenties to induce a flight of capital from the rural areas to the urban areas, basically to Mexico City.

Industrialization, the other major component of the Mexican development model, which is by far its dominant element, provided the "pull" or attraction forces that likewise brought about the MAMC's extraordinary economic and demographic growth. The driving motor of the Porfirio Díaz economy, external trade based on mining and agriculture, suffered a strong decline in the twenties as a result of the continuously strained international relations of the revolutionary regimes with the industrial powers. An inward-oriented economic policy took the place of the former Porfirian emphasis on external trade. A national policy to develop a domestic industry that would in stages replace imports with national products was adopted by the post revolutionary regimes in the twenties and was given the highest priority in government policy and budgetary assignments, as a way out from the straits of economic and political dependence. The creation of a large internal consumer and labor market in the heart of the country, large enough to absorb its output, was seen as a key element to achieving import-substitution industrialization.[6]

The agricultural investment capitals that fled the rural areas in the twenties found new opportunities in Mexico City in the expansion of existing manufacturing industries such as iron, steel, cement, paper, textiles, shoes, beer, tobacco, soap, sugar refining, and flour milling. These industries were all expanded from their pre-revolutionary base,[7] while new enterprises--among them metal goods, window glass, and foodstuffs--developed during this time.[8]

The industrialization drive became thus intimately linked to a set of associated strategies such as protectionism and centralization. Import substitution required that the national market be protected against external competition by a system of tariffs and import permits.[9] It also required that the growth of the capital city be stimulated, as a means of achieving the domestic and labor market necessary for indigenous industrialization.[10] Centralization became a most important support strategy of the economic and political model which dominated the government policies until quite recently.

The capital city became the main site of the national industrialization program. Mexico City's potential consumer market was relied upon as the cornerstone of this import substitution model, and the growth of that market was seen as crucial to its success. There was an awareness of the importance and great value of the externalities that the capital could provide to industrial development through its financial and commercial services, proximity to government, geographic centrality, and transportation. The agglomeration economies that were already available there to individual firms, industrial sectors, and industry as a whole, in addition to the economies of scale for public services, were recognized as a unique advantage to industrialization.

In terms of the amount of public expenditures, Mexico City became by far the most favored urban center in the nation. The location of public expenditures had important effects on the location of industry, "strongly affecting the attractiveness and even the viability of alternative industrial locations."[11]

The important motor road construction program that began to emerge in the twenties was totally dominated by the centralization principle. All the roads radiated from Mexico City which was to serve as the interconnecting central link of the system, joining the major urban centers with the capital. The road network followed the railway net, which was built in Porfirian times, strongly reinforcing the centralizing pull. The road network gave a far wider scope to the process of ending rural isolation, due to the much greater mobility and local penetration of motor roads, and literally paved the way to the capital for hundreds of thousands of isolated villagers.[12]

Crucial as Mexico City was in the economic model, it was even more central in the political model adopted by the post-revolutionary regimes. The model urgently sought to strengthen the central government, and as a consequence, Mexico City's primacy. To counter the local military caudillos who acquired great power during the revolution, threatening not only the central government but the political stability and unity of the country, a political model of an even stronger central government than in the historical past was adopted. This was based on an all powerful President overseeing a centralized government.

A corporative political structure representing the factors of production—peasants, workers and employers—was instituted in 1929. This structure gave the state apparatus the power to mobilize great sectors of the population to legitimize its actions, while keeping full control over them through their group affiliation. The representatives of these groups became the channels of communication of the political system. The president's position as the supreme arbiter and decision maker was thus reinforced. This naturally led to the functional and physical concentration of decision making organs in the capital as well as dominance of the capital city—the seat of the corporate state. Concentration of economic and political power served the purpose of strengthening the position of the center vis-a-vis the periphery.[13]

Inward orientation was further strengthened in the thirties when exports practically collapsed as the depression brought down mining prices and wiped out the foreign agricultural export markets. The diminished significance of external trade and the restriction on imports induced a decline of the coastal seaport cities such as Veracruz and a clear turn towards inward development.[14]

World War II gave the Mexican import substitution industrialization model its once-in-a-lifetime opportunity. New unsuspected opportunities of external origin which were most favorable to the development of industry on the basis of import substitution appeared. For the first time in many years the national economy was sailing with the wind. The disruption of international trade that resulted from World-War II and the mobilization of the U.S. economy for the war effort coincided with an expanded domestic market for industrial goods which were not available from abroad.

At the same time increased foreign exchange earnings originating in the traditional agricultural export sector became available, helping to finance import-substitution industrialization. War shortages in the U.S. market which raised the prices of agricultural commodities assured Mexican producers a high rate of return on exported primary goods during the war and the immediate post-war period. This unexpected sustained boost from abroad and the great tariff protections against foreign competition which were imposed by the Mexican government for many decades to come, made it possible for industry, even after the war, to reach very high annual growth rates.[15]

Mexico City's role as the central pole of industrialization was reinforced with each additional round of growth. In addition to its intrinsic unrivaled natural advantages for industrial location, the policies which were adopted during World war II and which are still in effect today distorted prices in favor of the capital city, introducing a strong factor of urban bias. These subsidies followed the other geographic allocations of public social expenditures, such as higher education and health expenditures, that were historically associated with Mexico City. Different subsidies skewed the price balance in favor of Mexico City, greatly strengthening its "pull" potential, far beyond the industrial employment opportunities that were available there. The following examples give an indication of the bias towards Mexico City.

First, fiscal incentives were offered industries to establish themselves within the Distrito Federal until 1955. These incentives were abolished for tax income reasons unrelated to location policy, and were maintained across the political dividing line in the neighboring State of Mexico which took over the industrial promotion role from the DF.

Also, freight rates established by the government-owned railroads were the most favorable in the country, even for heavy bulky raw materials, inducing the location in the metropolitan area of even those type of industries that required heavy raw materials.[16]

Food subsidies on tortillas, bread, beans, cooking oil and a few other basic products were established for Mexico City in the late forties. The price of beans and corn flour for the tortilla industry, so important in the Mexican workers diet, was maintained at a lower level in the capital city than in the rest of the country. The powerful government company CEIMSA (the later day CONASUPO), imported, at a loss, the commodities it needed to insure that the price of the vital components of the workers diet be kept down in the city.[17]

Water and sewer prices in Mexico City received major subsidies, to be covered by taxes and interest-paying loans. The real cost of a cubic meter of water was calculated in recent years to be five times the amount paid by fiscal consumers, and the operational cost of the drainage system around thirteen times the charged price.[18]

Subsidized fuel and electricity were provided to the city inhabitants and diesel oil was supplied at subsidized prices to the city buses.[19]

Public transportation prices in Mexico City received major subsidies. The subway system only generated twenty percent of its revenues, and the trolleys barely nine percent of their budget.[20] According to 1983 data, the price of a subway ticket was one thirteenth of its real cost, while that of an electrical trolley only one fifteenth. Diesel oil was supplied at subsidized prices to the city buses which only generated twenty seven percent of their budget.

Property taxes were undervalued in the DF when compared to other states.[21]

Rent prices below $ MP300 were frozen in Mexico City by means of an emergency decree at the beginning of World War II which has never been repealed. Hundreds of thousands of tenants below this ceiling thus received almost rent-free apartments from property owners who were cast into the involuntary role of subsidizing rent prices.

To the previous list of urban bias policies assisting the industrialization drive one should add government's passive and active involvement in promoting the irregular settlements as a housing formula for the working force. Contrary to the uncritical view taken for granted in the literature which sees the phenomenon of the irregular settlements as an unintended incidental result of urban growth, much historic evidence confirms the contrary proposition that the irregular settlements were more by design than by default part of a strategy directly related to economic government policy. This view was also expressed in the opening statement of the Tlaxcala Conference on land use problems in 1981:[22]

> As industrial development by itself was too weak to propel the concentration of an ample supply of labor in the industrializing cities, the population was being attracted by the promise of a lot that could be serviced some day.

There can be little doubt that within the hierarchy of policies pursued by the post-revolutionary governments, import substitution industrialization stood at the very top as the guiding policy which dominated all the rest. This policy not only was backed by the private sector but also found its most ardent advocates in the political left.[23] The import substitution model still dominated during the fifties, sixties and seventies.[24] It went unchallenged as long as the national economy showed remarkable growth and its DNP doubled.[25]

Against this general backdrop of the primacy of the protectionist import substitution model, and the strategic role that the MAMC played in its implementation, one can gain a deeper comprehension of the nature of the planning policies such as growth control and decentralization that have been advocated in the capital during the last forty years. The different independent explanations for the failure of these policies that were offered by the authors mentioned in the first chapter are far better understood within this general framework. One can now understand why planning policies which headed in a contrary direction to the economic policies, such as was the case with decentralization or control of urban growth, could not have been allowed to interfere with them.

As long as the succeeding Mexican governments were fundamentally committed to the import substitution model, there was hardly any chance for physical decentralization to be taken seriously. The policy reversal with respect to the role of the capital in this model which seems to have taken place with the rise to power of Ernesto P. Uruchurtu as mayor of the DF (1953-66), who banned new developments and abolished tax advantages for new industry in the DF area, was not really intended to radically change the role of the MAMC within the industrialization scheme.

The Uruchurtu policies were rather of a local nature. They only redirected new industrial expansion and irregular settlement activity a few miles to the north to the nearby jurisdiction of the State of Mexico, as proven by the major developing activity that was transferred immediately to the neighboring state.[26] The fiscal incentives to new industry were reinstated in the State of Mexico, where the state government adopted a permissive irregular settlement policy. The neighboring area increasingly took over the role of the DF as the main pole of industrial location and irregular settlement activity in the MAMC. Nothing changed for the MAMC as a whole with respect to the centralization policy.

The MAMC rapidly became the unchallenged industrial capital of the nation. Significantly, the type of industrialization that took place in the capital as well as in the country in general, especially since the sixties, was of a capital-intensive as opposed to a labor-intensive nature. By 1970 MAMC markets accounted for more than fifty percent of the total national demand for industrial products and for fifty percent of the industrial national output, although they only employed thirty eight percent of the nation's total industrial employment force.[27] As a result of industry's limited capacity to provide employment, opportunities had to be sought in the tertiary service sector by the great numbers of job seekers, a sector typically correlated to low income, unsteady insecure employment and a lack of employment benefits.[28]

At the beginning of the seventies the country faced its first serious economic crisis after many decades of stable economic growth. The balance of payments became seriously imbalanced. Imports grew at a much faster rate than exports. The expansion of industrial production, pushed import demand higher, while prices of Mexico's main export, agricultural products, declined on the world markets.[29]

The period after 1970 did not, however witness a change in course with respect to the import substitution policy; to the contrary this period saw a major reinforcement of protectionism. New elements were introduced into government policy within this framework: a) a turn to exterior sources of finance to offset the payments imbalance, b) great expansion of the role of the state in the economy in search of greater power, and c) a major increase in government deficit spending [30]

Centralization soon reached a point where the question of whether the advantages of the economies of scale had been eaten away began to be discussed. The marginal cost of some of the public services and of social overhead capital construction became comparatively higher in Mexico City than elsewhere.[31] Water costs kept rising as a result of the increasingly greater distances (already 200 miles in 1970) and higher pumping lifts to the valley of Mexico required for growing consumption. Sewage and residual water disposal also became more costly as a deep sewer and powerful pump system had to be built (1964-) to discharge the sewage outside of the valley to protect the low-lying city from being flooded by its own effluent, as a result of the city's continuous sinking due to subsoil water extraction. Electric power became more costly than in other cities because to the increasing complexity of the distribution network, and lack of nearby hydroelectric or geothermal power. Building and land development costs were higher than in other areas because of the soft foundation ground of the valley, high earthquake risk, and higher land and labor costs.[32]

Interregional traffic, much of which passes through Mexico City, became severely congested and costlier.[33] Intraurban transportation costs rose because of the motor vehicle slow-down and increasing investments in the highway and underground systems. The city's debt increased from 1978 to 1982 thirteen times, and the city had to rely increasingly on external sources, foreign and federal, to finance its debt service and subsidies.[34]

The negative economic and non-economic externalities associated with urban growth began to become highly visible in the MAMC. The quality of life started to deteriorate increasingly as additional people, vehicles, industries, and buildings were being crowded into the MAMC. Irregular settlement activity became the dominant component of urban expansion. Atmospheric contamination originating from industries, vehicles, irregular settlements lacking sewers, and the erosion of the desiccated valley floor, took on critical dimensions. Underemployment, poverty and urban crime indices rose rapidly in many of its areas.[35] Nevertheless, for the economic policy makers the diseconomies were not seen as important enough to override the agglomeration economies of the MAMC.[36]

In what seemed to be a major departure of the centralization policy the López Portillo government (1976-82) postulated decentralization as the guiding principle of the National Urban Development Plan issued by its new ministery of urban planning (SAHOP). This change, however, occurred while adhering to its predecessors fundamental commitment to import substitution and protectionism. As a result, the actions of the economic ministries, pointed in the opposite direction of the decentralization plans, showing little sympathy for them.[37] The diseconomies of the MAMC were not yet seen important enough to override its agglomeration economies. Even the mitigation of the negative social and environmental impact of centralization was given a very low actual priority, in spite of occasional lip-service to its importance.[38] Thus it was possible—as one example of many—that at the same time that the decentralization plans were announced by SAHOP, and a program to relocate many government offices from the capital to provincial cities was officially adopted, PEMEX, the national petroleum company, was beginning the construction of its fifty five story megastructure in the heart of Mexico City.

The import substitution policy which engendered and favored centralization came under scrutiny and heavy attack for the first time during the de la Madrid period (1982-88). After fifty years of unchallenged reign of protectionism and import substitution, a reorientation in economic policy took place. This happened as the country's economy went into deep crisis with the fall in oil prices and a huge foreign debt that suffocated economic growth. Encouraged by international organizations such as the World Bank and the International Monetary Fund, an alternative to the protectionist model was suggested linking Mexico more openly to international free trade. The ground work of a new free-trade economic policy was laid by joining the group of nations that subscribe to the General Agreement on Tariffs and Trade (GATT) in August 1986.

Paradoxical as it may seem, the new economic policy was accompanied by a reversal on the urban decentralization policy that had been adopted by SAHOP (then renamed SEDUE). The viewpoint of the economic ministries which considered that the agglomeration economies of the MAMC were far too valuable to be dismantled or tampered with, won added favor with the incoming government and led to a reversal in the decentralization policies which were adopted by the previous López Portillo government. The unique role of the metropolitan market as the cornerstone of national economic development was again reaffirmed.[39]

The policy change from import substitution towards a free trade economy took a firm hold with President Salinas de Gortari who turned back the statist tide and began a major privatization program of government owned enterprises and banks (1989-94). By reducing tariffs, eliminating import licenses, opening up financial markets, and changing some of the rules on foreign investments he put in motion a process which is leading to the final dismantlement of the protectionist system. The negotiations with the USA and Canada to sign a free trade agreement, which are underway, should bring the reorientation to its climax.

Looking back at the record of the economic policies of the succeeding governments, one must come to the conclusion that planning policies such as decentralization and urban growth control which were inconsistent with the objectives of import substitution industrialization were moving against the current and stood no chance of success. As long as the succeeding Mexican governments were committed to the import substitution model, there was hardly any possibility for any policy interfering with its objectives to be taken seriously. The incongruence between urban and economic policy relegated the urban plans to the province of idle plan making. The change in economic policy, however, which is now taking place from import substitution towards free trade, opens new opportunities for a more congruent relationship between urban planning and economic policy.

Prospective Changes of the Economic Model and its Possible Impacts

An important question which arises at this juncture, concerns the probable effects that the new economic policy might produce in the future on the MAMC. It is sound to assume that a weakening of the economic pull forces of the MAMC can be expected, at least as far as industrial concentration is concerned, as the MAMC location loses importance due to the new policy. This is a cogent conclusion which Ian Scott already suggested in a World Bank study a decade ago:[41]

> Industrial concentration also has been closely linked with industrial protection. Were the importance of the domestic market decline and the economy to become more open and competitive, **a central location would have become less crucial**, and prospects for the development of domestic markets in the periphery would have improved. In particular, the growth of demand in the periphery would imply the achievement of new minimum threshold sizes for noncentral markets. Thus, because of changing geographic demand, the overwhelming advantages of proximity to a specific market may have become less important (p. 245).
>
> There seems to be a close relation between the historical process of growth led by import substitution in a protected market and the process of centralization. In the future, the growth of exports and of the manufacturing sector will be required for sustained economic growth. This implies a more open economic environment, featuring less protection and a gradual increase in the ratio of exports to GDP, which implies, in turn, a corresponding increase in the ratio of imports to GDP. Assuming that these trends emerged, the role of the central market would be modified by the increased importance of external markets, **and it would be less important in the future to locate a business in Mexico City in order to have the greatest access to the market**. Given this situation, and taking account of transport cost, **location in the periphery, close to points of exit for**

> **manufactured exports, would thus become more efficient from the standpoint of the national economy than location in Mexico City** (p. 248).

Although it would certainly be unrealistic to expect that the agglomeration economies of the MAMC will suddenly be discarded,[42] it can be safely assumed that the free trade policy will set in motion important changes in the location of industries and related services. A growing proportion of new industries and business firms will choose the northern border region while some established industries may also choose relocation from the MAMC. A pattern of industrial location will take shape which will be less centralized, with its gravity center moving to the north. Such a movement, however, will not only face the challenge of expanding urban infrastructure and housing for the growing labor force in areas unprepared for rapid expansion, but also of implementing an environmental protection policy. The alternative would be the replication of the MAMC model, with its grave social and environmental problems. Such a prospect would be particularly bleak for Monterrey which seems destined to play a major role in the new economic scheme due to its strategic location and economic base.

Another no less important question concerns the problems that the relocating industries will leave behind in the MAMC. The precedent of the exodus of manufacturing firms from the Northeast-U.S.A. to the sun-belt and to foreign countries in recent years, presents a possible scenario of what might occur. These industries left behind great numbers of unemployed industrial workers, greatly exacerbating the crisis of the inner cities in the North. Relocation and job retraining have been slow, involving a painful adjustment process for the displaced labor force and the cities.

Although one can safely assume that the "pull" forces of the MAMC will be weakened as the free trade policy is being implemented, assessment of the demographic impact of this policy is hardly a straightforward simple matter. Population growth in the MAMC has its own momentum. Natural reproduction in the MAMC will by itself remain a very powerful factor. With the birth rate eventually coming down, the natural growth rate will descend. That gain, however, will be partly offset by a reduction in the child-mortality figures which are still very high in Mexico, not only in the rural areas but in the poor urban neighborhoods as well.[43]

The main effect of the new policy can be expected on the migratory balance sheet. While the deteriorating environmental conditions in the MAMC and the growing economic opportunities in the north and other development poles should continue to induce an outgoing migratory current from the MAMC, as shown by the 1980-90 data, rural migration to the MAMC is far from over. The MAMC received in the period between 1985 and 1990 an average of 217,000 migrants per year, the equivalent of 600 new city dwellers every day.[43] There is no reason to assume that the "push" forces will suddenly cease to expel people from the rural areas. The rural areas of Mexico, as in most of the rest of the Third World, are still growing in absolute numbers, as a result of high population growth rates.[44] Neither is there a reason to assume that the urban bias subsidies which have skewed the balance in favor of the MAMC will disappear over night. Some urban bias subsidies that are inextricably linked to social and political stability will remain indefinitely. Most of them will probably outlast the current economic policy changes. Thus, it can be concluded that although the proportion of rural migrants diverted north will be increasing, the rest will continue to migrate to the MAMC. Rural migration will remain a quantitatively important factor in the MAMC migratory balance sheet.

It is therefore more accurate to conclude that whatever impact the economic policy changes can have on the future course of development of the capital, continuing demographic growth and local population redistribution in the region of the MAMC may be expected. Population growth will have to be faced at least for a good number of years to come. Even if population growth should slow down or stabilize sooner than expected, economic growth and local population redistribution are by themselves powerful factors that will bring about continuous urban expansion on the periphery.

Instead of speculating on its desirability it would wiser to center the discussion around the question of how to confront that reality. It would seem far more productive to rather concentrate on how to cope with population growth and internal redistribution trends instead of leaving the city's expansion process to chance decisions. One can hardly disagree on this point with Scott, who has been keen on emphasizing the importance of urban planning and management to provide for that growth: "Many of the great problems of this massive region will result from its spatial organization and the way it is governed. In some very important ways, the quality of life of its inhabitants will depend on how these problems are resolved and not on the size of its population."[45]

The regional decentralization plans to attain a more satisfactory territorial growth pattern in the future acquire new significance in this context. The vision of a integrated system of well communicated nucleated towns in the MAMC wider region, each clearly separated from its natural surroundings and provided with housing, business and employment opportunities, makes sense as a realistic alternative to present suburban sprawl. The establishment of such towns is intimately linked to the question of implementing land reserves for urban growth and environmental protection as prescribed by the Constitutional amendment of 1976, which in its turn is directly related to the discussion on the property rights of the *ejidos* in the urban periphery, the main providers of land for MAMC expansion, whose crucial role in the failure of the planning policies in general and the proliferation of irregular settlements in the MAMC is later to be discussed.

MAMC GROWTH: ITS SPATIAL DIMENSION

Physical expansion of a city can adopt two basic forms: increased human and building density or expansion of the urban area on the surrounding rural periphery. Both have taken place simultaneously in the MAMC to accomodate its exceptional demographic and economic growth. The policies followed in the MAMC until the present intensified building and human densities to their limits, specially in the DF. The MAMC average density is higher than that of any other European or American metropolitan area including New York, Paris, and London and is higher than Tokyo's, considered one of the highest urban densities in the world (table 3). Nevertheless, while the built-up area of the MAMC is more compact than that of any other comparable growing metropolitan area, expansion on its surrounding periphery has also taken place on a major scale in a largely random manner.

To understand the nature of the expansion process it is necessary to refer first to the physiographic unit to which the city belongs: the valley of Mexico. The valley is an upland basin 2,300 meters above sea level, enclosed by mountain chains on three of its sides which place important natural restrictions on the MAMC growth possibilities: the Sierra Nevada to the east, the Sierra del Ajusco to the south, and the Sierra de las Cruces to the west. The valley is only open on the north side to the municipalities of the State of Mexico.

The built-up area of the MAMC is located in the southwest of the valley, covering around thirteen percent of the valley's total area of 9,600 square kilometers. Part of the city occupies some of the lands left by the desiccated lakes that existed in its lower flatlands in previous centuries. The city has reached the western and southern mountain ranges, while in the northeast it spreads onto the inhospitable, barren, dusty bed of Lake Texcoco, where many

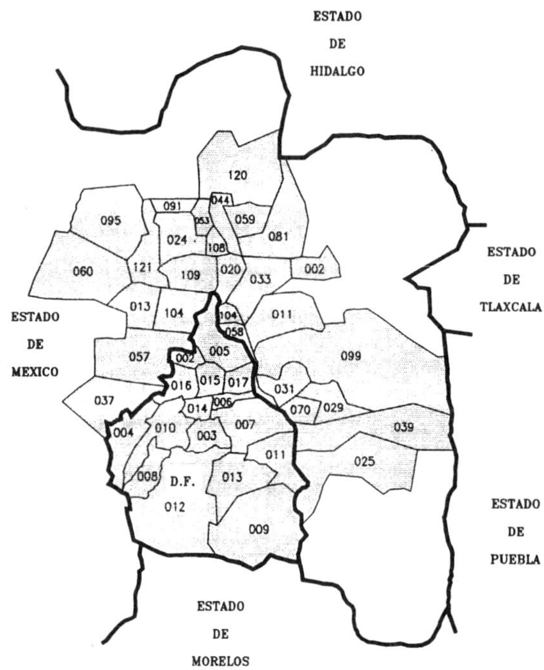

Figure 1. MAMC Political Divisions, 1990 (Source:INEGI, 1991)

DF DELEGACIONES

002 Azcapotzalco	003 Coyoacan	004 Cuajimalpa
005 G. A. Madero	006 Iztacalco	007 Iztapalapa
008 M. Contreras	009 Milpa Alta	010 Alvaro Obregón
011 Tlahuac	012 Tlalpan	013 Xochimilco
014 B. Juarez	015 Cuauhtemoc	016 M. Hidalgo
017 V. Carranza		

SM CONURBATED MUNICIPALITIES

104 Tlalnepantla	057 Naucalpan	033 Ecatepec
058 Netzahualcoyotl	013 A. de Zaragoza	020 Coacalco
039 Ixtapaluca	070 La Paz	109 Tultitlan
024 Cuautitlan R.R.	121 Cuautitlán Izcalli	029 Chicoloapan
037 Huixquilucan	081 Tecamac	025 Chalco
031 Chimalhuacan	060 Nicolas Romero	002 Acolman
011 Atenco	044 Jaltenco	053 M. Ocampo
059 Nextlalpan	091 Teoloyucan	095 Tepotzotlan
099 Texcoco	108 Tultepec	120 Zumpango

of the poorest irregular settlements are located. There is also a need to refer to the political jurisdictions on which the MAMC extends itself. It is necessary to distinguish between Mexico City proper; the Distrito Federal DF, the political division within which the city was originally contained; and the adjoining Estado de Mexico (State of Mexico, SM) into which it grew. The jurisdiction of Mexico City (Ciudad de México) was until 1970 limited to the core of the city that now corresponds to the four central *delegaciones* Cuauhtémoc, Miguél Hidalgo, Benito Juárez and Venustiano Carranza. Mexico City's administrative boundaries were extended in 1970 to coincide with those of the DF, and a new division into 16 administrative units known as *delegaciones,* instead of the previous 12 was adopted (fig.1).

The jurisdiction of Mexico City (Ciudad de México) was until 1970 limited to the core of the city that now corresponds to the four central *delegaciones* Cuauhtémoc, Miguél Hidalgo, Benito Juárez and Venustiano Carranza. Mexico City's administrative boundaries were extended in 1970 to coincide with those of the DF, and a new division into 16 administrative units known as *delegaciones,* instead of the previous 12 was adopted.The MAMC conurbation region, the wider geographic area that the Ministry of Programming and Budget SPP defined for planning purposes in 1983 in compliance with the Human Settlements Law, includes 53 municipalities of the State of Mexico, one municipality of the State of Hidalgo, and the 16 *delegaciones* of the Distrito Federal.

The city's built-up area was totally contained within the jurisdiction of the DF until the late forties. At the end of that decade, the urbanized area crossed the DF borders to the State of Mexico after President Avila Camacho (1940-46) created by decree the great industrial zone of Vallejo in the north of the DF in the *delegaciones* of Azcapotzalco and Gustavo A. Madero that border on the State of Mexico. Many of the newly created industrial plants were attracted to the Vallejo zone by the presence of the most important national railroad lines of Cd. Juarez and Laredo, and the availability of ample flat low cost land required for modern industry.[46] The establishment of new industries, and the growth of the population and the economy in the north of the DF produced important spatial impacts. Industrial development began to spill over into the neighboring SM municipalities of Tlalnepantla, Naucalpan, Ecatepec at the beginning of the fifties when the State of Mexico withdrew the ban on land sales that it kept between 1945 and 1951.

A powerful factor influencing the direction and character of the MAMC growth made its appearance in 1953 with the promulgation of a ban on new development in the DF by its mayor Ernesto P. Uruchurtu, and the cancellation of tax incentives to new industries in 1955.[47] The ban not only contributed to the increase of human and building density in the DF but also redirected urban growth to the neighboring State of Mexico. The State of Mexico, whose administrative and planning efficiency and resources were far below those of the DF, was interested in increasing its meager fiscal base and in commercializing its great extensions of land—many of them of marginal quality—in the near vicinity of the city. It did this by establishing tax incentives to new industries and by adopting a de-facto policy of unrestricted territorial expansion.[48]

As a combined result of the DF hard policies and its own soft policies, the State of Mexico captured a growing share of the land development market. Irregular developments without services for the labor force and the growing number of poor migrants grew indiscriminately at a very fast pace very often on inadequate land. A land development code of high standards which was adopted in 1958 remained largely unenforced. Instead of slowing down the city's growth, Mayor Uruchurtu's ban simply helped to divert it to nearby jurisdictions of the State of Mexico, lowering both the level of quality and the direct short term costs of development to the public treasury.

Although the lack of uniform accepted criteria to define when an outlying area can be considered part of the MAMC may be the cause of some numerical discrepancies, little doubt remains on the fundamental characteristics of the MAMC expansion process. The trajectory of the MAMC territorial expansion wave is reflected in the changing population statistics of the SM municipalities and the aerophotographic images of the urbanized areas in the region. The sudden leap in the census population growth rates from minimal values reflecting slow natural growth to high annual values of 5% or more indicating intensive migration, reveal the historical period when the different municipalities became part of the metropolitan area.

By 1950, Tlalnepantla, Naucálpan, and Ecatepec could already be considered part of the MAMC.[49] Netzahualcóyotl, the main pole of irregular settlement activity in the MAMC, joined the MAMC during the 1950-60 decade, reaching an annual growth rate of 19.6%. Atizapan de Zaragoza, Coacalco, Ixtapaluca, La Paz, and Tultitlán can also be considered part of the MAMC since that decade considering their annual growth rates above 5%. In the sixties, rapid urbanization advanced further to another four SM municipalities: Cuautitlán Izcalli, Chicoloapan, Huixquilucan and Tecamac. In the seventies, Chalco, Chimalhuacán, Cuautitlan R. R., and Nicolas Romero joined the urbanization

race making a total of seventeen municipalities within the MAMC.[50]
Further expansion occurred during the recent 1980-90 decade. INEGI, the National Institute of Statistics, Geography and Informatics, who for the first time is publishing aggregate data for the MAMC, considers that another ten municipalities should now be included in the MAMC making a total of 27 municipalities. These are: Acolman, Atenco, Jaltenco, Melchor Ocampo, Nextlalpan, Teoloyucan, Tepotzotlan, Texcoco, Tultepec and Zumpango. At the present expansion rate the advancing MAMC may reach the State of Hidalgo to the north of the SM during the next decade as well as the state of Tlaxcala to the east.[51]

After urbanization crossed the border to the State of Mexico, the SM increasingly became the main growth front of the MAMC. Urbanization growth rates in the SM rapidly exceeded those of the DF. By the seventies, 60% of total population increase during the decade took place in the SM. The changes of the 1980-90 decade were even more dramatic: the DF population ceased to grow altogether and was even reduced by 595,000 inhabitants, while the SM population grew by 1,917,000 people (tables 3 and 4). As a result of the process, the populations and urbanized areas of the two MAMC jurisdictions have almost reached parity. The State of Mexico municipalities already contained by 1990, 45% of the total population and 48% of the total urbanized area of the MAMC (table 3).

While the DF as a whole experienced substantial population loss, there are important variations in this respect within its boundaries. Population loss is limited to the core. The *delegaciones* Cuauhtémoc, Juarez, Hidalgo, Iztacalco, Venustiano Carranza, Gustavo A. Madero, and Azcapotzalco lost population at annual rates varying between 1.7% and 3.0% in the 1980-90 decade. One highly saturated SM municipality in the immediate vicinity of the core, Tlalnepantla, also registered population losses at annual rates of 1% during the last decade (table 5).

The southern periphery of the DF, on the contrary, continued to gain population and to increase its density and urbanized area. Coyoacán, Cuajimalpa, Contreras, Milpa Alta, Tláhuac, Tlalpan, Xochimilco and Iztapalapa grew at annual rates between 0.7% and 3.4% during the same period (table 5). Urbanization in these *delegaciones* continued to expand on the last remaining rural and forest reserve in the DF in a random and irregular manner. At least 222 sq. km. of the DF rural reserve were devoured in the 20 years between 1970 and 1990, while heavy damage was being inflicted on the environment, particularly on the aquifer recharge areas. Some of the DF core *delegaciones* mentioned began to show population losses since the sixties, as a result of their saturation,

Table 3. Population and Extension (sq. km.) of the MAMC Between 1940 and 1990

Year	Population	% of MAMC	Urbanized Area	% of MAMC	Density p/ha	Migrants %
1940						
DF	1,553,000	100.0	117.5	100.0	132.1	73.5
MAMC	1,553,000	100.0	117.5	100.0	132.1	73.5
1950						
DF	3,050,000	97.6	228.6	88.8	133.0	64.5
SM[a]	74,000	2.4	28.8	11.2	25.7	n.a.
MAMC	3,112,000	100.0	257.4	100.0	121.4	51.2
1960						
DF	4,871,000	93.0	310.5	76.1	156.9	41.9
SM[b]	365,000	7.0	97.4	23.9	37.5	85.1
MAMC	5,236,000	100.0	407.9	100.0	128.3	38.3
1970						
DF	6,874,000	78.4	459.1	63.8	149.7	34.7
SM[c]	1,891,000	21.6	260.2	36.2	72.6	88.8
MAMC	8,765,000	100.0	719.3	100.0	121.8	43.2
1980						
DF	8,831,000	64.3	638.0	56.3	138.5	21.4
SM[d]	4,895,000	35.7	495.2	43.7	98.8	n.a.
MAMC	13,726,000	100.0	1133.2	100.0	121.1	n.a.
1990						
DF	8,236,000	54.7	681.3	52.2	120.9	24.5
SM[e]	6,812,000	45.3	624.1	47.8	117.9	39.9
MAMC	15,048,000	100.0	1305.4	100.0	119.5	31.1
2000						
MAMC	16,622,000	100.0	1500.1	100.0	110.8	-
2010						
MAMC	18,361,000	100.0	1723.8	100.0	106.5	-

Sources: INEGI. VI, VII, VIII, IX, X, XI Censos de Población y Vivienda; Anuario Estadistico del DF, 1984. (a) Includes 3 municipalities; (b) 9; (c)13; (d) 17; (e) 27 .

Table 4. Population and Urbanized Area Increases (sq. km.) Between 1940 and 1990 in the MAMC

Period	Population Increase	% of MAMC	Growth Rate	Area Increase	% of MAMC	Growth Rate
1940-50						
DF	1,497,000	95.3	7.0	111.1	79.6	6.9
SM	74,000	4.7	-	28.8	20.4	-
MAMC	1,571,000	100.0	7.2	139.9	100.0	8.2
1950-60						
DF	1,821,000	84.9	4.8	81.9	54.4	3.1
SM	324,000	15.1	17.3	68.6	45.6	12.9
MAMC	2,145,000	100.0	5.3	150.5	100.0	4.7
1960-70						
DF	2,003,000	56.8	3.5	148.6	52.2	4.0
SM	1,526,000	43.2	17.9	162.8	47.3	10.3
MAMC	3,529,000	100.0	5.3	311.4	100.0	5.8
1970-80						
DF	1,957,000	39.4	2.5	178.9	43.2	3.3
SM	3,004,000	60.6	10.0	235.0	56.7	6.6
MAMC	4,961,000	100.0	4.6	413.9	100.0	4.6
1980-90						
DF	- 595,000	- 45.0	- 0.7	43.3	25.1	0.7
SM	1,917,000	145.0	3.4	128.9	74.9	2.4
MAMC	1,322,000	100.0	1.0	172.2	100.0	1.4
1990-2000						
MAMC	1,574,000	100.0	1.0	204.7	100.0	1.4
2000-2010						
MAMC	1,734,000	100.0	1.0	223.8	100.0	1.4
1950-90						
DF	5,186,000	49.5	2.5	452.7	43.2	2.8
SM	6,738,000	50.5	12.0	595.3	56.8	7.8
MAMC	11,924,000	100.0	4.1	1048.0	100.0	4.0

Sources: Based on Table 3

extensive land use changes from residential to business activities, and the deterioration of their quality of life. The outgoing population movement was greatly accelerated by the tragic 1985 earthquake which mainly hit the DF core. One can conjecture that this event will be perceived in the future as the turn point of the city's moment-ous population growth process which is slowing down sooner than expected.

The loss of people in the core and the move of people and activities to the expanding periphery of the MAMC are interelated phenomena. Of the 1,089,000 people which left the DF in the 1985-90 period, 549,000 relocated in the SM periphery while the rest moved to other states. DF local migration thus contributed around one third (28.6%) of the two million population increase in the SM metropolitan area during the last decade (tables 4 and 5). This is reflected in the unrelenting expansion of the MAMC on the farther out municipalities (table 3). The advance of the MAMC on 10 new SM municipalities in the last decade added 465,000 additional people to the MAMC statistics in 1990 (table 5).

Table 5. Population Growth in the DF *Delegaciones* and SM Municipalities of the MAMC (1980-1990)

Jurisdiction	Population 1980	Population 1990	Population Change	Growth Rate
DF	**8,831,079**	**8,235,744**	**- 595,335**	**- 0.7**
B. Juarez	544,882	407,811	- 137,071	- 2.9
Cuauhtemoc	814,983	595,960	- 219,023	- 3.0
M. Hidalgo	543,062	406,868	- 136,194	- 2.8
V. Carranza	692,896	519,628	- 173,268	- 2.5
Azcapotzalco	601,524	474,688	- 126,836	- 2.4
G. A. Madero	1,513,360	1,268,068	- 245,292	- 1.7
Iztacalco	570,377	448,322	- 122,055	- 2.4
Coyoacan	597,129	640,066	42,937	0.7
Cuajimalpa	91,200	119,669	28,469	2.7
Iztapalapa	1,262,354	1,490,499	228,145	1.7
M. Contreras	173,105	195,041	21,936	1.2
Milpa Alta	53,616	63,654	10,038	1.7
Alvaro Obregón	639,213	642,753	3,540	0.0
Tlahuac	146,923	206,700	59,777	3.4
Tlalpan	368,974	484,866	115,892	2.8
Xochimilco	217,481	271,151	53,670	2.2
SM Total	**4,894,575**	**6,811,941**	**1,917,366**	**3.4**

Tlalnepantla[a]	778,173	702,807	- 75,366	- 1.0
Naucalpan	730,170	786,551	56,381	0.7
Ecatepec	784,507	1,218,135	433,628	4.5
Netzahualcoyotl[b]	1,341,230	1,256,115	- 85,115	- 0.6
A. de Zaragoza	202,248	315,192	112,944	4.5
Coacalco	97,353	152,082	54,729	4.6
Ixtapaluca	77,862	137,357	59,495	5.8
La Paz	99,436	134,782	35,346	3.1
Tultitlan	136,829	246,464	109,635	6.0
Cuautitlán Izcalli[c]	173,754	326,750	152,996	6.5
Chicoloapan	27,354	57,306	29,952	7.7
Huixquilucan	78,149	131,926	53,777	5.4
Tecamac	84,129	123,218	39,089	3.9
Chalco[d]	78,393	282,940	204,547	13.7
Chimalhuacan	61,816	242,317	180,501	14.7
Cuautitlan R.R.	30,527	48,858	18,331	4.8
Nicolas Romero	112,645	184,134	71,489	5.0
Acolman[e]	-	43,276	43,276	-
Atenco	-	21,219	21,219	-
Jaltenco	-	22,803	22,803	-
Melchor Ocampo	-	26,154	26,154	-
Nextlalpan	-	10,840	10,840	-
Teoloyucan	-	41,964	41,964	-
Tepotzotlan	-	39,647	39,647	-
Texcoco	-	140,368	140,368	-
Tultepec	-	47,323	47,323	-
Zumpango	-	71,413	71,413	-
TOTAL	**13,725,654**	**15,047,685**	**1,322,031**	**1.0**

Sources: INEGI . X and XI Censo de Población y Vivienda , 1980; (AMCM, 1990).
a) Included in MAMC in 1950; b) in 1960; c) in 1970; d) in 1980; e) in 1990

A fundamental characteristic of the movement to the periphery of the MAMC is its being largely dominated by informal urbanization. It differs from suburbanization in the metropolis of the developed world by the predominant role of the poor in this movement, and the irregular character of their settlements. In a recurring cycle, the poor in the older SM peripheral settlements who can not afford its rising costs are now being forced out to the more distant frontier. The Chimalhuacan and Chalco municipalities are at present the most active irregular settlement front in the SM with average annual population growth rates of 14.7% and 13.7% (table 5). They are the most recent example of informal urbanization that dominates the MAMC expansion process on the periphery.

The Irregular Settlements

The irregular settlements are central to the problems facing the MAMC. Created without official approval and occupied by their residents long before their infrastructure is built and services are provided these settlements are probably the main symptom of the planning system's failure to guide urban development. It is necessary to look more closely at their historical origins and development in the MAMC to approach the discussion concerning the role played by the *ejido* property rights in the proliferation of these settlements.

The enormous growth of the irregular settlements on the urban periphery has its origin in the crisis of the traditional low income form of housing—the *vecindad*—which is the oldest type of low income housing in Mexico City. The *vecindad* is a type of rental unit going back to colonial times which is mainly found in the old core of the city. It typically consists of many overcrowded dilapidated dwellings, often sharing sanitary services, surrounding one or various courtyards, in old, sometimes historical buildings, which have gone through multiple subdivisions and adaptations.[52]

Until the forties many of the migrants coming to the city found shelter in the old *vecindades* of the central core. Demand for low income housing began to exceed supply when rents below $300 MP were frozen during World War II by an emergency decree promulgated in 1942. The *vecindades* lost their economic appeal for landlords and real estate investors. At about the same time unrealistically high building and land development standards in the DF, unaffordable to all except the upper income sector, were adopted.[53] The supply of new *vecindades* for the poor and even for the lower middle class ceased altogether.

The rent freeze, which was never repealed exerted a far-reaching influence on the central core by anchoring much of the lower income population benefiting from the frozen rents. The decay of the historical center and of many historical buildings followed.[54] Some parts of the central core of Mexico City went through rapid changes in its land use pattern as higher income residents moved to the periphery and other non-residential activities took their place. By the late forties the existing center city*vecindades* were saturated and stopped absorbing any additional population.[55] The core could no longer absorb the incoming migrants and the natural growth of the low income sector. The migrants joined the general movement to the periphery where they created many of the so-called irregular settlements.

A geographic differentiation in terms of income took place as the population moved towards the periphery. It evolved along lines defined by the earlier urbanization of the Porfirian era: the growing middle class and the rich overwhelmingly moved to the west and the southwest of the DF to well developed subdivisions, while the poor moved to the east and the north to illegal settlements in the SM without services, many of them on the inhospitable lands of the Texcoco lake bed. This differentiation was not only related to the varying land values and environmental conditions that prevailed in the different directions of the periphery, but also to an important degree to the tenure status and tenure problems of land in the periphery.

Thus developed the spatial pattern that characterizes the city today: a center in which there is a concentration of bureaucratic, commercial and service activities, surrounded by a horseshoe of intensely populated, low income, dilapidated slum neighborhoods such as Tepito, Guerrero, and Peralvillo, where the rent freeze largely paralyzed change; a northern area dominated by intensive industrial development mixed with low income housing for the working population; an eastern highly impoverished irregular settlement area on the inhospitable grounds of the desiccated lakes; a western area already reaching the hills, dominated by middle and upper income residential uses, mixed with some commercial uses, and a southern area sharing common characteristics with the western side but with a mixture of pharmaceutical plants and irregular settlements.[56]

While the rent freeze protected the existing tenants and favored industrialization by keeping down the cost of living for many workers, the housing market for low income people broke down on the supply side. As wages lagged behind rising land and construction costs, many low income people could not afford even the least expensive formal housing accommodations. The irregular settlements on the urban periphery were seen as the way out. By 1952 there were already more than three hundred irregular settlements in the DF alone, occupying 23.5% of the city's total area and containing fourteen percent of its population.[57]

The irregular settlements came within the reach of low income people because they lacked almost all attributes of urbanization that could increase their cost: property titles, urban design contracts, development permits, infrastructure, and public services. These costs were further reduced by building on marginal land unsuited to other uses. The settlers began by building temporary structures, upgrading them slowly as time and economic resources permitted. Left to themselves to find surrogates for the missing services, they bought water in retail quantities from water trucks or human carriers, dug primitive latrine pits for sewage disposal, hooked-up to the nearest electric power lines, and took hour-long walks to the nearest markets and highways to endlessly wait for overcrowded, infrequent buses. At no moment, however, were they free enough from their dependent relationship to be able to mobilize their own resources for full-fledged service provision, as property titles and service provision remained the sole prerogative of the state.

Looking at the irregular settlements from the viewpoint of the settlers, researchers have investigated the reasons why the irregular settlements in the urban periphery were favored over other forms of housing—even when other forms were available. John F.C. Turner (1972) found that many of the settlers chose this alternative because of their reluctance to assume the burdens of regular rent and mortgage payments associated with public housing. Self-built housing offered them instead a rent-free alternative which could be upgraded, keeping pace with their fluctuating income possibilities. R. Kemper (1977) pointed out that self-built housing in the irregular settlements responded to the settlers' wish to flee the cramped conditions of the inner city *vecindades*, replicating the spaciousness of their earlier rural home-sites, while allowing them the opportunity to use their slowly improving dwelling as a savings bank. Susan Eckstein (1977), found that the settlers valued highly the privacy and tranquility of the outlying *colonias* in contrast to the daily quarrels and gossip that are part of the crowded inner city*vecindades*.

It must be said, however, that valid as these reasons may often be for the low income population, self-built housing in the peripheral settlements was almost universally the only possible alternative available. This was particularly true for people lacking permanent employment, or whose income was below the minimum wage or even below some substantially higher limit. Even when social housing programs began to be implemented in the sixties for certain groups of the population, these families did not qualify—much less could they access the extremely scarce credits from the private banks. With neither the formal housing market nor government providing housing for them, they were forced to turn to the informal market and accept the hardships of living in an irregular settlement.

The Irregular Settlements and Public Policy

The government or its agents patronized the creation of many of the settlements and took a tolerant, permissive attitude to the formation of many others. A specific account of the role of the government in creating the irregular settlements in the earlier days was formulated by a spokesman of the settler movement (Unión de Colonias Populares del Valle de México) at the Tlaxcala Conference, who brought out the artful role of government in the irregular settlement movement:[58]

> Many irregular settlements, *colonias proletarias,* were created in Mexico City by the government which stepped in and distributed lots in sites that were beginning to be settled, to win over some sectors of the masses and appear to be the creator of these settlements. Thus were established the *colonias* Alvaro Obregón, 20 de Noviembre, Mártires de Río Blanco, Progresista and Azteca in 1938, as were the *colonias* Revolución, Damián Carmona, Francisco Villa, Río del Consulado, Emiliano Zapata and Gertrudis Sánchez in 1939. The 50,000 urban lots that were thus distributed by the governments of Cárdenas and Avila Camacho according to the testimony of a leader of those days illustrate the scope of the distribution policy.

The active role played by the DF city government in promoting irregular settlements in the period before 1953, is well documented by various authors.[59] Mayor Javier Rojo Gómez (1940-46), backed by President Avila Camacho, is recognized as the promoter of large-scale irregular settlement activity, through what came to be known in those days as *paracaidismo*, or invasion of land. He also carried out expropriations to regularize the legal situation of some of these settlements and distributed their lots under the *patrimonio familiar* regime, which made it illegal for the beneficiaries to market or mortgage them.[60] His successor, Fernando Casas Alemán (1946-52), is likewise recognized as having actively backed large-scale *paracaidismo* activity during his administration, but refrained from any further expropriations.[61]

As the State of Mexico increasingly took over the role of the DF as the main pole of industrial location in the MAMC, irregular settlement activity increased. Government involvement in promoting the irregular settlements in the SM became even more direct after 1952:[62]

> The State of Mexico's Junta Federal de Mejoramiento Cívico, Moral y Material, which was created in 1952, had negligible resources to help resolve the problems faced by the settlements in the east of the city. Its real purpose was to stimulate the proliferation of subdivisions and, through its liason with the CNOP wing of the PRI, to gain political and electoral control over the area.

Even within the DF itself, irregular settlement activity did not cease altogether, although government participation there became less direct. In spite of Mayor Uruchurtu's official ban on any type of development (1953-1966),[63] irregular settlements were established to the south of the DF on the land of *ejidos* in the delegaciones of Coyoacán, Ixtapalapa, Ixtacalco, Xochimilco, and Tláhuac. Many of the people displaced by public works, were also relocated there. Quoting the same source mentioned earlier:[64]

> During the fifties and sixties the government party (PRI) became the organizer and promoter of the *colonias proletarias* with the tolerance of the authorities. The mechanisms of political clientism turned the settlers in one of the most servile sectors of Mexican corporativism. In spite of the land development regulations adopted by the city government (DDF), and confronting a situation of legal anachronism in land use questions, many new irregular settlements arose in the DF such as Ajusco, Ruiz Cortines, Díaz Ordaz, Isidro Fabela, Santa Ursula Coapa, Santa Martha Acatitla just to mention a few of them.

The early seventies in particular brought about many invasions of *ejido* land in the south of the DF. The largest invasion known took place in 1971, when 20,000 well organized people occupied overnight land that belonged to the Copilco el Alto *ejido* in Coyoacan. Land invasions continued with the covert support of politicians and government officials.[65]

Their greatest growth, however, took place in the sixties and seventies in the State of Mexico to the north and northeast, in the municipalities of Tlalnepantla, Naucalpan, Ecatepec and Netzahualcóyotl. Great extensions of undeveloped land were illegally subdivided and enormous numbers of lots were sold. By 1980 there were already around thirteen hundred irregular settlements in the MAMC, containing as much as sixty percent of the total population. In the State of Mexico, its most active growth front, it is estimated that almost 4.5 million people, 63.1% of the total population of the municipalities are living in them.[66]

A policy in favor of the irregular settlements seemed to fit in well within the interests of the state. The irregular settlements were seen by the government as an escape valve to social tension building up in the rural areas and as a means to facilitate industrial development. Their political value as an instrument to win popular support in the urban areas for the political establishment in no way escaped political strategists.[67] Government could win popularity among the urban settlers at no direct immediate cost to the public treasury and without assuming the embarrassing role of officially promoting illegal actions.

The lack of property titles, infrastructure, and urban services in the irregular settlements obtainable only from government, gave the government a powerful long-term lever over the irregular settlements. Commitments for satisfying the settlers needs for urban services could be avoided and postponed for decades to come and traded slowly for political loyalty. The political value of the irregular settlements as a means to replicate in the cities the relationship of clientism and unending heavy dependence on government already existing in the rural *ejidos* was recognized very early on. A policy favorable to the irregular settlements certainly had many short-range political benefits.

The rent freeze and the irregular settlements, the two pillars of the housing policy for the masses pursued by the government since 1942, were capitalized on politically by the government to draw into its ranks the growing urban poor and lower middle class. As early as 1941 Mayor Rojo Gomez approved a law, the *Reglamento de las Asociaciones Pro Mejoramiento de las Colonias del DF* which granted legal recognition to settler organizations independent of the legal validity of their claims on the land they held.[68] Only one organization could represent a settlement according to that law, giving the DF government the discretional power to choose among conflicting groups. By this means the DF government acquired great political power over the settlers. To gain recognition the settler organizations were required to join the government party and to participate in its activities.[69]

The government party created a third political sector within its structure in 1943, the so called *Sector Popular,* to draw in within its ranks all those left out of the peasant (CNC) and labor sectors (CTM). Organized under the name of National Confederation of Popular Organizations, the CNOP assembled the rapidly growing irregular settlements population demanding land, titles and services and the tenant organizations opposing rent increases and favoring the rent freeze. The CNOP thus succeeded in seizing the representation of an important constituency in the periphery of the rapidly growing urban centers. Also included within the CNOP were the lower government bureaucracy, the teachers federation and the self employed.[70]

The government could now operate its control of the irregular settlers through coopted leaders capable of maintaining their demands within the limits acceptable by government. The Office of Colonias Proletarias within the city government and the CNOP within the governing party, were placed in a strategic, intermediary broker position between the higher government authorities and the settlers to carry out this delicate balancing act.[71]

Government could have responded to the modest demands of the lower income groups supplying enough vacant land in the urban periphery in areas with appropriate environmental characteristics, which would be easy to service in the future, allowing the residents to take care of their needs through self building. The government, however, did not design its settlement policy within the frameworks of legality and planning. The creation of the irregular settlements was left to chance decisions following the line of least resistance in a chaotic and disorderly manner.[72]

It is clear that the long range severe problems caused by accumulating service deficits for millions of people and promoting unplanned growth on untitled or vulnerable marginal land—often completely unserviceable or more costly to service in the future—were traded for short-term advantages. The long term impoverishing effects of dependence and the possibility of a breakdown of the political control mechanisms, were presumably not considered important enough. The thought that the bill would have to be paid some day was simply pushed aside.

Irregular Settlement Typology

Various types of irregular settlements came into being in response to low income people housing demand. Some were established by invasion of squatters without any ownership claim on the land, others by promoters illegally subdividing and selling land, and still others by the *ejidos* in the periphery using as a cover the law allowing them to establish residential villages for their members in their *zona urbano ejidal* (ZUE). Whatever their differences they all share in common their lack of infrastructure and services and the lack of recognized property titles for the settlers.[73]

Invasion Settlements

Invasion settlements are those established by squatters without any original rightful property claim on the land. The *colonias paracaidistas* and *ciudades perdidas* are two varieties within the invasion settlements category. The *colonias paracaidistas* are created in massive overnight surprise operations, from whence is derived their suggestive name, *colonias paracaidistas* "parachutists colonies", while in the *ciudades perdidas* "lost cities" squatting occurrs over an extended period of time without any plan.

As brought out in the literature, considerable strategic planning and organization entered in carrying out an invasion operation to create a *colonia paracaidista*. Streets and lots were mapped out in advance on primitive grids by the leadership, and the settlers were assigned their lots and provided with identity cards. An invasion operation to create a *colonia paracaidista* would normally have the covert backing of some powerful organization, political groups or figures within government. Sometimes, however, invasions were the result of an autonomous action by a group of settlers under independent leadership. In the later case their chances of success were smaller, unless they looked for a political sponsor.

Eviction was one of the risks that the *paracaidistas* had to face when the legal owners, private or public, undertook legal action to affirm their property rights. The success of the invasion operation depended greatly on the number of settlers that could be mobilized to avert the danger of eviction.[74] Unless eviction took place immediately, the settlers would move swiftly to obtain a court injunction and the legal struggle would continue indefinitely. While settlers were actually never evicted in former times, the pending eviction suit would be used by government to deny any petitions for services, until regularization—the final goal of the squatters—was agreed upon by the city authorities. However, in the meantime as soon as the organization of the settlers was granted official recognition and the organization joined the CNOP, the settlers would gain some quasi-legal status on the basis of what was called *derecho de posesión*, right of possession. While the original squatters gained their right to a lot free, any new settlers coming in after the recognition of the settler organization would have to buy their lot from the leaders of the *colonia* in which an informal land market would begin to function fully.

Although no land was immune to invasion, it was *ejido* land more than any other type that offered the invaders the best chance of success, where they could easily outnumber the *ejidatarios*, even though the *ejidatarios* would present resistance, and physical conflict would ensue.[75]

The *Ciudades perdidas,* the other variant of squatter settlements, would not display the strict geometric pattern of the *colonias paracaidistas* but would instead present a disorderly agglomeration pattern without any dividing property boundary lines. In these high density agglomerations of flimsy one-room shacks one could count twenty or more families living in an area equivalent to a minimal city lot. These settlements have arisen even in the vicinity of middle class or upper class subdivisions, filling-in non-urbanized marginal land unsuitable for formal development, such as abandoned sand mines, disused open-cast clay mines, ravines, gullies, lava formations, and mountain slopes. Sites with legal ownership in dispute, or unproductive *ejido* and *comunal* property would also be susceptible to this type of squatting.

Unlike the *colonias paracaidistas* regularization of tenure rights would rarely occur in the *ciudades perdidas,* which would be kept in a continuous state of insecurity. Introduction of services and construction of streets would be very costly and technically difficult. Like displaced-persons camps, the *ciudades perdidas* would not show any improvement over time in their environmental and housing conditions.

Another important distinction must be made here between real invasion settlements and pseudo-invasion settlements which really belong in other categories. In the last case, an invasion operation would occur with the connivance of the owners and even as a result of their initiative, in order to open up the land for urbanization through a fait accompli in situations where the authorities would not allow development. Many pseudo-invasions on private and *ejido* land, real and simulated, were prompted when Mayor Uruchurtu attempted to restrict land development activity within the limits of the DF through his development ban (1953-66). Such pseudo-invasions nullified the ban's effectiveness to a large measure within the DF. Since the inauguration of the López Portillo government in 1976, government policy has been not to favor and suppress invasion operations in the DF area.

Irregular Developer Subdivisions

The irregular developer subdivisions are urban settlements created by real estate companies or landowners who, without compliance with the requirements of the land development codes to introduce the minimal prescribed services, subdivide and sell lots in the urban periphery at low prices. The development codes prohibit the sale of lots before services are introduced in order to prevent from leaving the burden on the government. Lacking the necessary official permits the development companies are unable to conduct their business within the formal market and resort to informal contracting with their clients. These settlements are included in the loosely defined category of *colonias proletarias* the so called proletarian settlements mentioned in the literature.

The land on which the illegal developer's subdivisions are located is either of marginal quality, unfit for other more profitable purposes, or is often subject to serious land tenure problems that block its access to the formal market and legal development activity. It is quite common that the status of the land as private, *ejido*, or public property is in serious doubt, and that developers original property rights are not established in the public registry, or are under litigation.

Nevertheless, a more stable and permanent situation exists in the irregular developer subdivisions when compared to the invasion settlements. Undisturbed by eviction fears the settlers will normally show greater confidence in the future of their *colonia* in their long quest for regularization. Although the settlers are buyers, not invaders, due to non-compliance of the subdivisions with the development legal code and the often undefined tenure situation of the land, they will nevertheless be unable to obtain individual property titles, even when their payments are completed.[76]

The developer settlements will typically show a primitive design pattern never going beyond a monotonous rectangular grid. Houses will be laid out on endlessly repeated rectangular unpaved streets delineated by a few curbstones and served by some electric power lines with a few street hydrants for collective water supply. A visitor will not find these settlements in their early decades, with their dusty, unpaved streets, self-built makeshift dwellings, lack of vegetation, heaps of garbage and street dogs, much more appealing than the invasion settlements. Their physical form is defined generally by separate individual one-family lots, where the owner family is often joined by relatives or tenants living in adjacent structures. Sometimes they engage in minor livestock raising and obtain some cash by subletting part of the lot to poorer settlers.[77]

Most of the irregular developers' subdivisions were established in the State of Mexico. The SM government actively supported this type of development activity.[78] Landowners and private companies in close combination with political powerful figures detected untapped business opportunities from the early forties in the State of Mexico and acquired great extensions of marginal land. The commercialization of these settlements has been shown to have produced great profit margins for many developers, particularly in the sixties. Although lot prices had to be within reach of the poor, the profits made were still great.[79] Government support continued at least until the end of the sixties, when the companies ran into serious difficulties as a result of the general payments strike that the settlers initiated against them, demanding more services and compliance with their legal obligations. The payments strike and the federal government's clamp-down on the development companies seemingly brought to an end the activity of these companies in the SM.

The most conspicuous example of this type of subdivision, described in more detail in the case studies cited in chapter V, was created by important real estate companies and finance banks in the municipality of Netzahualcóyotl. This huge agglomeration of subdivisions was established on the dried out marshlands of Lake Texcoco, an area unfit for any other purpose but wetland rehabilitation. The land also faced serious tenure problems resulting from conflicting claims. The claimants were thousands of *comuneros,* the federal government, and private owners who received endowments of land in the twenties. These barren, inhospitable lands were used by development companies to create the greatest agglomeration of irregular settlements in the continent, in which most services were lacking. It now houses a population of nearly 1.5 million people living in one and two-story makeshift constructions, with a density of around 263 people per hectare.

***Ejido* Settlements**

The *ejido* settlements are informal subdivisions lacking official approval and services created by the *ejidos* on their non marketable land in the vicinity of urban areas. At least until 1970 such *ejido* settlements were initiated on the basis of a law included in the agrarian code in 1954[80] which allowed *ejido* communities in the rural areas to allocate a fraction of their agricultural land to establish a residential village, a *zona urbano ejidal* (ZUE) for their members, while the main fraction of the land was to continue in agricultural use. The *zona urbano ejidal*, clearly intended in the agrarian legislation for rural areas, is defined in the law as the site where the *ejido* community can concentrate their houses and community installations in a compact layout, in order to "facilitate the task of providing the peasants with the benefits of potable water, electricity, and in general the public services necessary to protect their health and promote progress in the rural environment."

Within the *zona urbano ejidal,* each member family of an *ejido* community is entitled according to the agrarian law, to a parcel of up to 2,500 square meters (26,910 square feet), free of cost, to build their house. These parcels are distributed by means of a lottery. The process frees the residential parcels from its restricted *ejido* status, changing them to private property, which the owners are free to sell or lease. The approval of a ZUE petition is an highly complicated and very slow legal process that takes many years, involving as many as 138 different steps.[81] A similar process is required to approve the expansion of an already existing ZUE, when it becomes necessary as a result of the *ejido's* own population growth.

The agrarian law also allowed until 1971 the *ejido* community to sell or rent excess parcels that are available in the ZUE at officially assessed commercial prices, to non-*ejido* members called *avecindados* "who wish to settle within the *ejido* villages to cooperate with the effort to develop them."[82] The law indicated that the proceeds of the sales should go to the common fund of the *ejido*, to be spent on "collective service works." The suburban *ejidos* were thus provided in this clause with an escape route to circumvent the fundamental non-marketability restriction of *ejido* land. The petition to establish or expand their ZUEs was used by the *ejidos* in the vicinity of the MAMC as a means to undertake subdivision of their land while their petition was being processed by the agrarian authorities, long before any services would be introduced.

During the early sixties the government actually granted ownership titles, *Certificados de Derecho a Solar Urbano*, to *avecindados* in the MAMC duly signed by the President of the Republic as was the case of the San Juan Tlilhuaca *ejido* where a a few thousand certificates were awarded to settlers.[83] The notion of using the ZUE figure to create *colonias populares* had gained such momentum with the assistance of the agrarian authorities that according to data by Ann Varley as many as 91% of the *ejidos* in the metropolitan area had applied to the Departamento Agrario for the approval of a ZUE by 1970. While the *ejidos* could only access the informal market with its lower prices they nevertheless had a strong incentive to subdivide and sell to settlers willing to take the hardships and risks of this market as will be later discussed in greater detail.[84] Once the subdivision process started it was made extensive to the totality of the *ejido* not merely to the original designated fraction. Even the best type of agricultural land would thus be subdivided and sold by the individual *ejidatarios* and their elected authorities.

The result in a environmental sense was in no way different than in the invasion and company settlements. The urban government would refuse services to these settlements over many years and decades as long as the ZUE approval was absent. Lack of infrastructure and services in addition to the uncertainty on the outcome of the ZUE approval process acted as a powerful obstacle to improvement. The settlers in the *ejido* subdivisions were left to wait and hope for the recognition of their *colonias* as ZUEs and their incorporation into the city and its service networks while the *ejidos* continued their subdivision and sales activity. Not the threat of eviction, but recognition of the ZUE status of their settlements and introduction of services was their main concern.

A radical change in policy with respect to the ZUE approach took place with the Echeverria administration (1970-76) at the same time that company sponsored subdivisions of irregular settlements were brought to an end. The *avecindado* clause was amended and limited in 1971[85] to only such candidates "that are engaged in an occupation which is useful to the *ejido* community," a definition which obviously disqualified as *avecindados* most of the settlers in the ZUE settlements with one stroke. When regularization of the *ejido* settlements was undertaken in response to the settler movement of 1970, the ZUE procedure was declared to be not applicable and expropriation by government and resale to the settlers recognized as the only acceptable legitimate route. A later amendment of the agrarian code incorporated an explicit negative reference to the ZUE approach that was condoned for 19 years stating: "It will be required in every case, to justify the effective need of establishing or expanding the ZUE, to satisfy preferentially the needs of the *ejidatarios* themselves, and not of those of the nearby towns or cities."[86] A closer look at the policy change in later chapters will reveal its rationale and objectives.

The dismissal of the ZUE option in 1973 in no way meant that informal subdivision in the *ejidos* had come to a close. After company sponsored development was terminated in 1970 and invasions began to suppressed since 1976, the *ejidos* in the periphery became the main and almost exclusive source supplying land for irregular settlers. Irregular settlement activity became almost synonymous with *ejido* urbanization. Informal urbanization is advancing on the MAMC periphery through the subdivision of the *ejidos* in the DF and State of Mexico periphery as intensely as before, supplying an apparently boundless market of lower income people eager to acquire a low cost parcel in the urban periphery.

Regularization of the Irregular Settlements

The irregular settlements began to generate serious social conflicts and unrest in the MAMC in the late sixties. Typical of this period, when an intense power struggle developed before the upcoming presidential elections, active power groups made efforts to capitalize on deeply-felt demands for property titles and services by the millions of irregular settlers. The tame Frankenstein—to use Gilbert's expression—began to turn restless, as settlers, developers, and *ejidos* became involved in growing confrontations. The irregular settlements policy faced its first major political crisis after more than thirty years of successful political maneuvering.

Conflicts between settlers and *ejidatarios* broke out in 21 of the 87 *ejidos*. At the same time, a massive payments strike was declared in Ciudad Netzahualcóyotl against the developer companies. This signaled the beginning of a crisis in company-sponsored irregular subdivision activity. The payments strike against the development companies in Netzahualcóyotl was followed by a government clamp down on these companies, bringing to an end their feverish business activity in the State of Mexico for over twenty years.[87]

Recognizing that land property titles were the common goal of almost all the settlers and a major obstacle to the improvement of the irregular settlements, a program to regularize tenure in them was announced by the incoming Echeverria government in 1970. Tenure regularization was perceived as a first step leading to the introduction of services by the DF and SM governments and the incorporation of the settlements into the tax base. Described in greater detail in the corresponding case study in chapter IV, Netzahualcóyotl became the first test site of the regularization programs that were later applied in other settlements. A special government trust fund was created to carry out regularization in Netzahualcóyotl which took over the responsibilites of the companies.

The varying nature of the tenure problem in the other irregular settlements led to the creation of federal and local government agencies to handle the different situations of land tenure regularization. The DF government founded FIDEURBE for regularization and development of *ejido* and private property in the DF on June 1, 1973.[88] The federal government established CORETT, Comité de Regularización de la Tenencia de la Tierra, for the regularization of settlements on *ejido* land on August 20, 1973.[89] AURIS, created earlier by the SM government for general urban planning and development purposes, was charged with some of the land regularization and commercialization programs in the State of Mexico. DART, an agency for the regularization of irregular settlements on private property, was created by the DF government four years later in 1977.

As the overwhelming majority of the irregular settlements were located on *ejido* land, CORETT became the most active and important of all these agencies. Although national in its scope, almost all of CORETT's regularization programs were located within the MAMC, particularly within the DF jurisdiction.[90] The 1974 charter authorized CORETT to apply the powers of the Federal government to expropriate occupied *ejido* land in order to regularize the irregular settlements and provide the settlers with private property titles.

A less advertised objective of the regularization program was to generate profits for the public sector. Aware of the high profits that developer companies had obtained from low income land subdivision, the incoming government wished to channel these profits to the public sector. The government intended to achieve this objective by combining the regularization programs with land subdivision activity. It gave CORETT authorization to use its expropriatory powers to acquire vacant *ejido* land to anticipate urban expansión which it could commercialize or use to carry out new subdivisions. Expropriation of the still vacant portions of the suburban *ejidos* for profitable commercial purposes became part of the regularization programs. The agency was turned into an instrument to extract financial resources and profits from the regularization of urbanized *ejido* land, and from the subdivision and commercialization of vacant *ejido* land in the urban periphery for the agrarian sector. The financial resources were to be channeled to FONAFE, the agrarian ministry's development fund, for the economic development of the *ejidos* in the rest of the country.

The *ejidatarios* responded to the CORETT and FIDEURBE regularization programs by stepping up the pace of the illegal sales of land. They anticipated the regularization programs by preemptively subdividing and selling as fast as they could as much *ejido* land as possible to potential settlers and private buyers, greatly expanding the irregular settlements area. Why this was so is a subject of discussion in the literature and will be dealt with in the next chapter. However, authors Gilbert and Ward have interpreted this effect of the regularization programs, as a result of the incentive that the additional regularization payment presented to the *ejidatarios* to profit doubly. According to this view the *ejidatarios* would first profit from illegally selling the land to settlers, and again when receiving compensation through the regularization program.[91] This view is challenged in the next chapter as presenting an oversimplified picture of the situation that leads to inadequate understanding of the real issues involved.

CORETT's instrumental role as the agency in charge of the regularization programs of the irregular settlements on *ejido* land has become permanent. However, the clause allowing CORETT to acquire vacant *ejido* land to integrate land reserves was revoked in 1979.[92] The responsability for establishing such reserves was transfered instead to the newly created Ministry of Human Settlements (1976-82) which was assigned in the Human Settlements Law the responsibility for creating and managing such reserves. A special department for that purpose, the Dirección de Reservas Territoriales was established in the Ministry of Human Settlements to implement what became known as the *Programa de Reservas Territoriales.*[93]

Although establishment of land reserves for urban expansion or for housing purposes had to wait until 1984 to be explicitly recognized in the agrarian legislation as a valid ground for expropriation, it was nonetheless assumed that this was a justified public interest purpose. The many cases where *ejido* land had been expropriated for public works, universities, sport installations and other public purposes by presidential decree was considered solid ground for applying the same precedent to solve the needs of the public housing agencies or the expansion needs of the population centers in general.

The establishment of a government organization at the highest ministerial level was seen as the missing link in implementing the program. The problems that such specialized organizations as CORETT, FIDEURBE and AURIS had encountered in taking over *ejido* land for development and commercialization would finally be overcome by a land reserve program of a ministry specially created for urban planning purposes. It was assumed that as owner of urban land reserves the government would be in a position to actually implement its urban development plans and anticipate urban growth.

Actual results turned out quite different. The land reserve program intensified the illegal sales of *ejido* land even to a larger extent than the regularization and reserve programs of CORETT and FIDEURBE. The *ejidatarios* in the peripheries began to sell land frantically. Efforts by the Human Settlements Ministry to hold off the process demanding stiffer sanctions against the *ejidatarios* or to accelerate takeover of *ejido* land, now encountered the opposition of the Ministry of the Agrarian Reform, the agency in which lay final authority to carry out the expropriations of *ejido* land. The Ministry of the Agrarian Reform and CORETT took a position in favor of regularization as opposed to preventive expropriation of vacant land before development occurred.[94] The situation which developed is appropriately described by author Azuela de la Cueva:[95]

> As a consequence, a rift has developed since then between the sector in the public administration in charge of promoting the constitution of reserves before the lands are occupied, and the agrarian sector in charge of the regularization of land tenure where the land has already been occupied by settlements. But this is not a simple rift but rather a racing contest. Every hectare which is not expropriated in time for the constitution of reserves for a city is sold by means of informal mechanisms and becomes the subject of regularization. The crucial point in this case is that who decides the pace of the expropriations—the Ministry of the Agrarian Reform on whom CORETT depends—is not responsible for fulfilling the objectives of the urban development sector. The result is predictable: the expropriations for regularization purposes are processed much faster than those where the objective is to establish land reserves.

As a direct result, expansion of the irregular settlements on the *ejidos* in the MAMC periphery reached its highest historical level after the Ministry of Human Settlements and the DF government were made responsible for its implementation nullifying the objectives of the reserve program and shattering any illusions on the possibility of harnessing the land resources of the *ejidos* for planned growth.

Significant conclusions can already be drawn up in this chapter on the evolution of government policy regarding the irregular settlements in the MAMC. It should have become abundantly clear that the irregular settlements are still playing a key role in solving the pressing housing needs of the low income people and that government policy has not been able to substitute this solution for any other one. Public housing has not yet become a real alternative as will be shown in the ensuing section. Some of the means to implement the irregular settlements formula may have changed, such as land invasions or informal development by private companies which are not encouraged anymore, but government policy remains fundamentally tied to this housing formula as in the historical past. The *ejidos* in the urban peripheries are now being assigned a dominant role in the formation of the irregular settlements which other agents shared in the past with them. This role is given a measure of legitimacy by the regularization programs which the agrarian establishment carries out at some indefinite time after they are created.

Regularization can be seen as a relative improvement over the earlier policy adopted at the threshold of the forties which allowed the proliferation of the settlements leaving them in legal limbo for decades, but it can not substitute for planned growth. The belated actions of regularization can seldom remedy the harm done by irregular growth and can not take the place of timely expansion that anticipates growth. Adopting regularization as a surrogate to planned growth as it has been done until now constitutes an admission that government is incapable of leading and managing the urban expansion process.

The physical, economic and social impacts of the regularization programs have been the subject of discussion. While there is general agreement that regularization has acted as a catalyst which unchained a process of physical improvement and the release of pent-up social energy and entrepreneurship where it has been implemented, some authors have come up with the contention that regularization is against the interests of the poor.[96] They argue that the regularization programs trigger a process which displaces the poorer settlers to more distant locations in the receding periphery where they begin a new informal settlement cycle.

They point out that the monetary outlays required to obtain title and services through regularization add to the land's cost. Incorporation of the land and buildings to the tax base increases these costs further. Some settlers face the higher costs of holding property in the regularized settlements by selling. The poorest settlers drop out altogether at the very beginning of the race unable to afford the regularization payments. Others do so later, while some others face these higher costs by subdividing, renting or sharing their real estate, increasing the densities in the settlements. When services are eventually introduced and land values and taxes rise again, rents experience an upward pressure, forcing out the poorest tenants. People in higher economic brackets move into the area, taking the places of some of the original settlers.

The concern of these authors for the negative effects of regularization on the poor is certainly justified. It should lead to a realization that the irregular settlements should not be seen as an opportunity to generate profits for the agrarian sector. However, their arguments in favor of maintaining the status quo or of countering the market effects of the programs by applying a form of non-marketable property known as *patrimonio familiar* which would not allow their owners to sell the regularized properties, is less defensible. The greatest opposition to such proposals has always come from the settlers themselves whose most cherished goal is the acquisition of fee simple property rights.

The side effects of regularization can be expected in a market situation. They may seen similar to those observed in gentrifying neighborhoods in a developed metropolis, or for that matter, in any area were population mobility is not artificially blocked and the value of property rises. However, in the case of the irregular settlements the main beneficiaries of the increases in property values are the settler families themselves. By introducing their land into the formal primary market, regularization produces increases in the market value of their land and buildings, benefiting them. The great differentials between the secondary and the primary market prices that exist prior to regularization diminish, and price levels tend to equalize. They rise with the rest of society instead of stagnating when others prosper.

PUBLIC HOUSING AS AN ALTERNATIVE TO THE IRREGULAR SETTLEMENTS

It is now necessary to examine the question whether any other formula within government policy such as public housing was available to diminish the need for the irregular settlements. An historical overview leaves little doubt that public housing did not provide such an alternative in the crucial period of MAMC expansion, although public housing was relatively early incorporated into political discourse as a formula to remedy market failure. Not only because such a task was beyond the budgetary means of the government, but also because public housing policy was not really designed for that purpose. It still does not provide such an alternative, although important adjustments in public housing policy have occurred in more recent times which point towards meaningful quantitative and qualitative changes in this respect.

While the housing problem in Mexico, as in any other developing country, is ultimately rooted in the income limitations of a majority of the population (table 8), this factor taken by itself would be of little value in explaining the evolution of public housing policy over time. The limitations on the supply side of the housing market in Mexico are of crucial importance. It is also necessary to focus on the effects of the government's own more general policies. An analysis of public housing in Mexico will show that many of its public housing programs became necessary to counter effects induced by the government's own political, monetary, fiscal or urban policies.

The housing market on the supply side was hardly equipped to face the demands of a rapidly growing population. The lack of bank mortgages to allow home buyers to acquire property was one of the major structural chronic weaknesses of the Mexican real estate market, which historically limited housing supply at all levels until the present. The periodic currency devaluations and chronic inflation dwindled away the value of the fixed-rate mortgage loans provided in Mexican currency. The private banking system, as a result, largely refrained from investing its scarce funds in home mortgage lending because of the rapid loss of value of such loans. These funds rather went to short term loans for industry and commerce. Only people with sufficient cash to pay its full price could access the market to acquire an home. All others, even with high incomes, could only satisfy their housing demands by renting.

The high cost of the legal transfer of property in Mexico, which could exceed 10% of assessed value, was another factor that made housing more costly for all sectors. This factor also contributed to tenure illegality in the expanding periphery. The notary fees and taxes involved in the act of transference of real estate property have been since olden times an important source of income for the state treasuries and the DF government which they protected zealously. The public housing agencies looked for ways of getting around this obstacle to reduce housing costs by various devices. Only recently are these costs being reexamined by government to bring them in line with internationally accepted standards.

The building industry was still a slow small-scale operation of craftsmen which had not yet incorporated modern production and management techniques. The difficulties in obtaining building permits even for one-family houses, were present at all levels, but were particularly frustrating to the modest home builder. Last, but not least, the obstacles faced by the land developers in obtaining permits for subdivisions, due in part to the high costs of expanding the water and sewage systems for the MAMC, contributed to the quantitative and qualitative inadequacies of the supply of housing in the MAMC.

Public Housing Policy From 1947 to 1970

One can hardly speak of a public housing policy before 1947. Until the end of the Avila Camacho administration in 1946 such a policy can only be described as an effort to ease the lack of mortgage credits for a very small number of government employees which it sought to provide with mortgages to build or buy their own housing. The Dirección de Pensiones, the government employee's pension fund which evolved into what today is the ISSSTE, began in 1925 to provide individual mortgage loans to a very limited number of fortunate affiliates. From 1925 to 1946 it extended such credits to a total of 9,600 affiliates.[97] Home ownership was a very limited option for most of the population. The rest had to turn to the rental market to take care of their housing needs.

The rental housing market faced its first crisis, at least as far as the low income people is concerned, with the rent freeze decree of 1941. In a situation of rapid population growth in the MAMC and unsatisfied demand, government began its first building program to increase rental housing supply for its own employees during the presidential period of Miguél Aleman (1946-52). Using the pension funds of the government bureaucracy as investment capital in housing, the Dirección de Pensiones inaugurated the first rental housing project in Mexico in 1949, the Multifamiliar Miguel Alemán.

The large-scale public housing projects in postwar Europe and the USA provided the model for the housing units, while the "trickle down" theory provided the policy rationale of building these middle class rental housing units for the bureaucracy. The economies of scale of the projects were presented as an argument to justify this type over conventional rental housing, although many affiliates objected strongly to them. As it turned out later, maintenance costs in the projects were substantially higher than in conventional apartment buildings. The large rental projects built by government for the bureaucracy soon superseded individual mortgage lending and became solidly entrenched in the public sector, not the least because of the patronage opportunities which they provided in the building and occupation stages.

The rental projects reached their climax during the fifties. A total of thirteen large rental projects in the MAMC were built by the Dirección de Pensiones. Other government organizations followed the early example of the Dirección de Pensiones and also began to build rental housing units for their affiliates. IMSS, the Instituto Mexicano del Seguro Social, the health care and pension fund organization created by President Avila Camacho (1940-46) for the workers, adopted the idea of investing its pension funds in housing during the López Mateos administration (1958-64). Six large-scale rental housing projects for a minority of chosen beneficiaries were built by IMSS during this period, among them the Unidad Independencia and the Unidad Santa Fe in the southern DF.

Government policy rapidly undermined its own long-term ability to maintain and increase investment capacity in public rental housing. As a result of its policy of not raising rents in the projects, a practice obviously linked to its populist political stand on maintaining the rent-freeze decree and winning popular political support, a situation was rapidly reached where not only were utility margins for reinvestment eaten up by inflation, but maintenance costs in the units could not be covered anymore. The large scale design of the projects also made maintenance costs higher than in conventional isolated buildings. The units rapidly became a burden instead of an asset to the agencies. Even before private sector investors left the rental housing field, the public sector ceased investing in rental housing almost completely[98] and began a program of condominium conversion of its rental units, a process which often confronted strong tenant opposition.

The government also established during the forties the Banco Nacional Hipotecario Urbano y de Obras Públicas, (BNHUOPSA, later BANOBRAS) which was the first government organization to enter the housing market as developer of land and builder of middle class housing for sale in the MAMC. BNHUOPSA searched for a way to go around the obstacle of the high costs of property transfer and applied in its units a transitional form of tenure established in the law in 1946,[99] known as the *Certificado de Participación Inmobiliaria* (CPI). A buyer of a CPI acquired transferable property rights to a specific dwelling unit without being required to pay the property-transfer taxes which would be postponed to the end of his mortgage payment plan. Property under this legal regime would be held in trust by the mortgage bank, until the buyers finished to pay their mortgages. Transfer of the building and apartments to the permanent property regime chosen by the certificate holders would then take place.

Another important legal development that produced a far reaching effect on the housing market is the law of condominium property adopted in 1954 during the administration of President Ruíz Cortines (1952-58).[100] This law allowed people to become private owners of individual dwellings within a larger building (vertical condominium) or a lot held in common (horizontal condominium), while sharing property on all the common elements. It was intended to stimulate greater investment in home building in a situation of rapidly rising prices of land, as was occurring in the MAMC and the larger cities. Although its effects were slow in the beginning, it soon became a very important factor in the housing market which contributed much in modifying the ratio of owners to renters.

The first step to expand public housing policy beyond the narrow limits of the bureaucracy to the poor was taken by President Adolfo Ruíz Cortines with the creation of the Instituto Nacional de la Vivienda INV (the later INDECO now extinct) in 1956. Such an initiative was hardly more than a symbolic gesture at the time. The *Herradura de Tugurios* surrounding the downtown center with its run-down*vecindades* and the rapidly growing irregular settlements in the periphery were singled out as the two most important areas in need of assistance. While the INV raised the level of awareness on the nature of the housing problem for the poor, and clearly articulated in its policy proposals the crucial importance of providing land and essential infrastructure as well as technical assistance and low-cost materials to the settlers in the periphery at a time when such an approach was not yet popular with the international development organizations, the agency lacked any significant budget means to implement its proposals.[101] More important, the idea of providing adequate land and technical assistance to the self-builders was still alien to the urban authorities. Neither did such an approach fit in well with the dominant housing philosophy which understood public housing exclusively in terms of large projects and building contracts.

Tenant ownership instead of tenant renting became the preferred housing formula adopted by government after it began its retreat from rental projects in the early sixties. A greater availability of funds, internal and external, allowed the governments of López Matéos (1958-64) and Díaz Ordaz (1964-70) to expand its capacity of supplying mortgage credits to prospective buyers of apartments in public housing projects. The housing project concept was combined during this period with the idea of urban redevelopment which had become popular in the industrialized nations during the fifties. A vast redevelopment project coordinated by the Ministry of the Treasury, Nonoalco Tlaltelolco, was built during the López Matéos administration on 75 hectares of cleared land previously occupied by railroad yards. The strategic location of the site between the major Insurgentes artery to the west and the newly opened northern extension of the Reforma boulevard to the east, lent this huge project its economic appeal. The Treasury invested reserves of the banking system through the Banco Nacional Hipotecario, BNHUOPSA in 86 buildings while the Dirección de Pensiones invested its pension funds in 16 others. Many of the buildings contained as many as 290 apartments each.

Although the Nonoalco Tlaltelolco project was originally planned as a rental housing unit, BNHUOPSA put the dwellings up for sale in 1964, while the Dirección de Pensiones held on to its original rental plan. The units were sold under the*Certificado de Participación Inmobiliaria* (CPI) regime. This decision turned out to be a most cumbersome one for the government in more than one respect. Twenty years later the government could not get the tenants to take over the administration of the project and after the 1985 earthquake had to foot the total reconstruction bill of the many destroyed or damaged buildings.[102]

A further important development in housing policy took place as a result of the Kennedy administration's offer to provide external aid to developing countries through the Alliance of Progress in 1963. The result was a program by the Mexican Treasury, the *Programa Financiero de la Vivienda* (PFV), which allowed the government to build housing projects in which the income threshold for aspiring candidates was lowered, although it was still considerably above the minimum wage. The Secretary of the Treasury, established a revolving fund FOVI using the external funds provided by the Alliance in combination with government and bank reserve funds to build *vivienda de interes social*, social interest housing for people with a stable income between three and six times the minimum salary capable of complying with the necessary mortgage payments.

While the quantitative balance until 1970 shows a steadily increasing participation of the public sector in housing, the numbers of dwellings built and their share in the total market was still small. In the 1950-60 decade the public sector as a whole built 62,000 dwellings in the whole country, 5.4% of the total national increase of 1,150,000 dwelling.[103] During the decade of the sixties that number rose to 175,000 dwellings, 9.3% of the national total increase of 1,877,000 dwellings (tables 6 and 7). However, more significant is the fact that although most of the public housing dwellings were built at that time in the MAMC they hardly provided an alternative to the irregular settlements. Public housing units, whether rental or for sale, were few in number and built for a fraction of the bureaucratic elite with steady employment and relatively high wages. Although the FOVI program increased supply since 1965 and addressed a wider audience, it really did not slow growth in the irregular settlements. The majority of the population within the lower income brackets or those holding unsteady jobs were completely left out of these programs. Public housing hardly provided any alternatives to them.

Public Housing Policy From 1970 to 1988

Government policy strongly affected the private rental market during two crucial periods, the early forties when rents below $MP 300 were frozen, and again in the early seventies when economic policy brought about a sharp decrease of private sector investment in rental housing. The same effect that the rent-freeze had in the forties on investments in *vecindades* became extensive to the rental market as a whole since the seventies. As the official bank interest rates soared as a result of internal and external lending and spending, rates of return of rental housing rapidly fell below the bank rates. The erosion of the return rates led very rapidly to the almost total extinction of new investment in rental housing since

Table 6. Number of Dwellings Built by the Public Sector in Mexico by Presidential Periods (1947-1990)

Presidential Periods	(1947-64)	(1965-70)	(1971-76)	(1977-82)	(1983-88)	(1989-90)	Total
(1925-1990) Pensiones-ISSSSTE-FOVISSSTE	45,302	1,300	27,027	56,628	92,658	161,461	277,173
(1947-1990) Pensiones Militares-FOVIMI-ISSFAM	1,100	0	1,566	2,369	2,549	969	8,553
(1947-1990) PEMEX	13,100	0	0	3,939	23,037	531	40,607
(1947-1990) BNHOPSA-BANOBRAS- FHP-FONHAPO	24,098	16,664	18,537	6,192	245,068	105,160	415,719
(1947-1964) IMSS	0	0	0	10,600	0	0	10,600
(1956-1990) INV-INDECO-AURIS-FIVIDESU	22,300	9,800	46,877	96,904	22,660	96,946	295,487
(1965-1990) FOVI-FOGA-Nationalized Banks	0	92,015	92,419	209,313	468,637	128,347	990,731
(1972-1990) INFONAVIT	0	0	101,448	262,889	414,206	161,461	940,004
(1986-1990) Earthquake Reconstruction Program	0	0	0	0	66,063	3,167	69,230
(1947-1990) Others	4,700	0	0	0	45,656	42,289	92,645
Public Sector Total (Includes Housing Support Actions Since 1983)	121,200	119,779	287,874	650,547	1,392,089	586,544	3,158,033
Public Sector, Finished Dwellings (National)	121,200	119,779	287,874	650,547	903,105	284,923	2,367,428
MAMC (% of Public Sector Total)	-	-	-		-	16.4 %	19.8%

Source: SEDUE 1990; CIHAC 90; VIII, IX, X, XI Censos de Población y Vivienda.

the early seventies. Private owners of rental buildings in the MAMC began to desinvest, converting their buildings into condominiums—wherever that was possible. While the conversion of rental units into condominiums was in itself a very positive development that allowed many people to become owners of their homes and facilitated the renovation of many buildings, the freed resources were not reinvested in housing. The rental stock in the MAMC suffered further losses by the collapse and deterioration of the old rent-freeze buildings in the core. The process of disappearence of rental housing accelerated in the more recent period. In the last ten years from 1980 to 1990 rental units dropped from 41.7% of the total stock to 25.6%. In the MAMC as a whole that share dropped to 21.9%.[104] Nationally, that percentage is even smaller, 14.6%, according to the 1990 census data. Public housing had to confront a much tougher job in order to make up for private sector desinvestment in rental housing and the rapid disappearence of the rental stock through deterioration.

Table 7. Total Increases of Dwellings in Mexico and the MAMC by Decades (1950-1990)

	<	Nationwide	>	< MAMC	>
Year	Population	All Dwellings	Public Housing ß	All Dwellings	Public Housing
1950	25,791,000	5,259,100	13,579	600,769	-
		(100%)	(0.3%)	(11.4%)	
1960	34,923,000	6,409,100	75,579	1,006,923	-
		(100%)	(1.2%)	(15.7%)	
1970	48,225,000	8,285,700	250,579	1,565,179	-
		(100%)	(3.0%)	(18.9%)	
1980	69,393,000	12,074,600	959,579	2,745,200	-
		(100%)	(7.9%)	(22.7%)	
1990	81,249,645	16,197,802	2,377,028	3,135,669	-
		(100%)	(14.7%)	(19.3%)	
Increase					
1951-60	9,132,000	1,150,000	62,000	406,154	-
		(100%)	(5.4%)	(35.3%)	
1961-70	13,302,000	1,877,000	175,000	558,256	-
		(100%)	(9.3%)	(27.9%)	
1971-80	21,168,000	3,789,000	709,000	1,180,021	162,120
		(100%)	(18.7%)	(31.1%)	(4.3%)
1981-90	11,847,000	4,123,202	1,417,449	390,469	225,556
		(100%)	(34.4%)	(9.4%)	(5.5%)

Sources: VII, VIII, IX, X, XI Censo de Pobación y Vivienda; SEDUE.
ß Only includes complete dwellings not diversified assistance actions.

However, this period is linked to a major increase in public housing as a result of the creation of INFONAVIT, a national housing fund for the work force, created under the same principle as IMSS, the public sector organization providing health care services and retirement pensions to the workers. INFONAVIT was established in 1972 through an amendment to article 123 of the Constitution which imposed a general 5% tax on the payroll of every business firm to provide housing for their employees.[105] Affiliated employees became entitled to low interest mortgage credits as these were becoming available.[106] The INFONAVIT charter made ownership the exclusive objective of its program.

INFONAVIT as the rest of organizations involved in public housing adopted a price policy which it upheld until 1987, by which the mortgage credits established at the moment of sale remained fixed until the end of the payment plan. It took a good number of years for the organization to realize that such fixed credits in a context of inflation and currency devaluations could only lead to its decapitalization. These factors reduced the real value of the outsdanding mortgages to pennies and led to the destruction of the capacity of the organization to grow financially and serve new groups of workers. A minority of affiliates was being subsidized by a majority waiting endlessly for their turn.

Provision of land for the public housing agencies became a major concern during this period. Various organizations became involved in this process such as INDECO (the former INV, 1973), AURIS (1969), BANOBRAS (the former BNHUOPSA), and CODEUR. The international development organization adopted during the seventies a policy favoring self-built housing and advocated involvement of the governments in its assistance. This had some limited effect on government policy in the irregular settlements, which led to the regularization programs and the establishment of construction materials parks through INDECO and AURIS for the first time. However, their recomendation of providing land in anticipation of development as an alternative to the irregular settlements was rejected because the government feared that it would add incentives to mass migration to the MAMC.

External and internal lending allowed the government to rapidly expand its housing investment programs during the 1970-80 decade. The annual public investment budget in housing multiplied five times from $MP 8,008.8 million (in fixed 1980 pesos) in the beginning of the decade to $MP 39,943.7 at the end.[107] The number of dwellings built in the country by the public sector increased from 175,000 units to 709,000 units, 18.7% of the total housing increase during the 1970-80 decade (tables 6 and 7). However, as public housing gained ground in the 1970-80 decade, private sector housing plummeted from an average 27.5% share to 16.5%. Self-built housing, on the other hand, the typical form of housing in the irregular settlements, remained at the same high level of around 65% of the total built during this entire period. The public housing gains were basically achieved at the expense of private sector housing.[108] Public housing grew at the expense of private sector activity without bringing about any substantial relief to those that needed it most—the self builders in the irregular settlements. In the MAMC the public sector as a whole (INFONAVIT included), produced at the peak of its activity (1973-1980), an annual average of 16,212 units in the MAMC, barely 13.7% of the annual dwelling increase in the MAMC during that decade.[109]

The decade of the eighties witnessed various momentous events and political changes which had far-reaching effects in a quantitative and qualitative sense on public housing policy. Especially significant in this respect are such events as the drop in world oil prices which threw the national economy into deep crisis, government's takeover of the private bank system at the end of the administration of President López Portillo (1986-82), and the 1985 earthquake which left many people homeless in the MAMC. In the critical period that followed after the nationalization of the private banks, President de la Madrid turned public housing into a motor of the economy to which he assigned 15% of the total government budget, raising the share of public housing from 0.84% of the GDP to 1.56%.

FONHAPO, the Fondo Nacional para la Habitación Popular was established in 1981 by the government to assist some of the lower income groups not covered by INFONAVIT or FOVISSSTE. The organization was assigned the task of providing partial support to people without formal employment, earning 2.5 times the minimum wage—an income requirement which still leaves out 65% of the population in the MAMC in 1990 (table 8). Instead of relying on the finished housing programs to indirectly diminish demand for self-built housing, FONHAPO was charged with the responsability of providing assistance to the self builders and thus directly improve that type of housing. A variety of programs to facilitate the building process of the self-builders and to help the slum-tenants were instituted, such as sites and services, building materials parks, credits for home building in stages, loans for tenants to acquire in ownership rent-frozen apartments (Casa Propia program) and credits for home improvements.

The earthquake that struck Mexico City on September 19, 1985 produced a major loss in the DF housing stock and created a critical housing situation for tens of thousands of people which became homeless overnight. Over seven thousand people were killed, 412 buildings collapsed, another 5,725 buildings were seriously damaged. Over 23,000 people had to live in street shelters for months while more than 30,000 doubled up with relatives or remained in their heavily damaged dwellings.[110] Major buildings of the Nonoalco Tlaltelolco compound collapsed and others were heavily damaged. Deteriorated *vecindades* and buildings in the poor *colonias* of the core caved in, leaving the tenants literally in the street.

Table 8. Individual Income Distribution by Minimum Wage Groups in the MAMC (1990)

Income Group as a % of Minimum Wage	% of Working Population in Each Income Group MAMC	DF	SM
0 to 100 %	19.2 %	20.7 %	17.0 %
100 to 200 %	41.7 %	39.7 %	44.4 %
200 to 300 %	15.8 %	15.4 %	16.3 %
300 to 500 %	10.7 %	11.0 %	10.3 %
500 to 1000 %	6.2 %	6.6 %	5.5 %
1000 and higher	3.4 %	3.5 %	3.3 %
Unspecified	3.0 %	3.0 %	3.0 %
Total	100.0%	100.0 %	100.0 %
Working Population	4,975,780	2,884,807	2,090,973
People/Household	4.8	4.6	5.1
Earners/Household	3.0	2.8	3.3
Official Daily Minimum Wage in July, 1990=$MP 10,080 ($US 3.5)			

Source: INEGI, XI Censo de Población y Vivienda, Area Metropolitana, 1990.

The government prudently decided to shy away from relocation schemes at the wrong moment and to rebuild on the same sites where the people had lived and kept their jobs. Three reconstruction programs with a total of 69,144 dwellings were carried out between 1986 and 1989: Renovación Habitacional Popular (48,000), Fase II (12,670) and Nonoalco Tlalteco (9,744). In a manner that can be assumed would have merited Jane Jacobs' and Herbert Ganz' approval, the government decided to rebuild an improved version of the former *vecindades* on the same sites for the same people, keeping alive the existent urban texture and community in the RHP program. The central patios and wide entrance gates, which were a major concern of the people, were preserved. The expropriation of 4,321 sites for the RHP program was carried out without major political turmoil. A loan from the World Bank was combined with fiscal resources to replace 42,090 *vecindad* dwellings and rehabilitate 6,610 existing ones in the RHP program. The program turned all the former tenants into owners of their modest (400 sq. ft.) but much improved dwellings at a cost of 3.21 million pesos per unit ($US 5,246 at the 1986 exchange rate). The attempts of political groups to establish restricted non-marketable forms of tenure were rejected for full property rights that allowed the owners freedom to dispose of their property according to their best judgement.

In order to avoid costly long-term reponsibilities, government acted within the principle of transfering ownership and management responsibilities to the tenants RHP as soon as possible. A short-period payment plan based on minimal prices was established, eliminating any mortgages which could have yielded higher rates of return, but involved long-term responsibilities. The special ad-hoc agency, Renovación Habitacional Popular which was created to carry out the program was to be liquidated as soon as building and transfer of the properties would have been carried out. Another significant aspect was the approval of an amendment to the Law of Condominiums,[111] creating the *Condominio Vecinal* which could be applied to hardship situations as these. The amendment granted discretionary legal authority to the DF government to exempt these *condominios vacinales* from building requirements that they could not meet, such as parking spaces, as well as to forgo property and transfer taxe in the case of the original tenants.

A very important development in housing policy took place during 1987 in INFONAVIT, when it was finally realized that the fixed mortgage credits were not only inimical to the organization's long-term investment capacity, but were providing incentives to their affiliates of selling the units to make a quick profit. INFONAVIT decided to adopt the principle of indexation and tie the monetary value of the mortgage loans to the minimum salary, to provide greater protection to the financial resources of the organization and discourage speculative behavior. The price of the INFONAVIT dwellings would thus be determined in multiples of the minimum salary, reducing the gap between the commercial value and the buying price at any given moment. A proposal to apply the same principle to rentals was presented by SEDUE in its 1988 housing plan, in order to establish a more stable and equitable relationship between owners and tenants.[112]

The housing balance of the 1980-90 decade shows that the public sector has made great advances towards becoming the dominant power in housing building. The annual public investment budget in housing doubled from $MP 39,943.7 million to $MP 81,654.0, from 1880 to 1990 (fixed 1980 pesos).[113] The number of dwellings built nationally by the public sector also doubled from 709,000 units, 18.7% of the total increase in the 1970-80 decade to 1,417,449 units, 34.4% of the total 1880-90 housing stock increase. In the MAMC that dominance is even greater, due largely to the official reconstruction programs after the 1985 earthquake and the slow-down of population growth in the MAMC (tables 6 and 7).

While an increase of such a magnitude should lead to a growing diminution of self-construction and irregular settlement activity, especially in a context of slower population growth, this last housing form is still the dominant growth factor in housing in Mexico. According to a 1991 CIHAC estimate, of the 634,875 dwelling units presumably built during 1990 in the country, 52.4% (330,000 units) belong to the self-construction builders, 160,000 units (25.2%) to the public sector and 144,875 units (22.8%) to the private sector.[114] The conclusion can thus be drawn that although there have occurred very importante quantitative and qualitative changes in public housing policy in recent years, public housing is still far from providing an alternative to the many millions of people in the lower income strata that fall outside of its parameters.

---

[1] INEGI, XI Censo de Población y Vivienda, 1990: Cuadro 4.2.

[2] (Grindle, 1986).

[3] (Pommier, 1982: p. 346-347).

[4] (Cornelius, 1975: p. 21-26; Kemper, 1977; Muñoz & Oliveira, 1973; Larrisa Adler Lomnitz, 1977).

[5] (Pommier, 1982: p. 346-347).

[6] (Scott, 1983: p. 44-45).

[7] Important industrial activity took place in Mexico City in Porfirian days (see Scott, 1983: p.34).

[8] (Adler, 1978).

[9] Solís, 1970: p. 173-184

[10] That intimate relationship is recognized in the literature: There seems to be a close relationship between the historical process of industrial growth led by import substitution in a protected market and the process of centralization (Scott,1983).

[11] (Scott, 1983: p. 111-115; Pommier, 1982: p. 355-357).

[12] (Scott, 1983: p. 39-44).

[13] (Aguilar Camín & Meyer, 1990).

[14] Ibid.

[15] There is a wide literature on this period and subject which includes such authors as Sanford Mosk, Clark Reynolds, Leopoldo Solís, René Villarreal, Timothy King, Dwight

S. Brothers, and Roger D. Hansen.

[16] (Scott, 1983: p. 245).

[17] The company imported, at a loss, the commodities it needed to insure that the vital components of the workers diet: tortillas, bread, beans, cooking oil and a few other basic products were available to the working class in the D.F. at a low and stable price. The price of beans, so important in the Mexican workers diet, was maintained at a lower level in the capital city than in the rest of the country. Subsidies were established on corn flour and dough for the tortilla industry (Fernandez y Fernandez & Acosta, 1961. Grindle, 1986).

[18] (Scott, 1983: p. 118; Campbell & Wilk, 1986: p. 312-313).

[19] (Scott, 1983: p. 242; Gilbert, 1982).

[20] (Campbell & Wilk, 1986: p. 312-313)

[21] (Scott, 1983: p. 118).

[22] (SMP, Sociedad Mexicana de Planificación, 1982: p. 9).

[23] VL Toledano, an ardent advocate of industrialization, understood this to be the most effective route to socialism.

[24] In the first stage of industrialization, (1924-1950), import substitution was devoted mainly to non-durable goods, in its second stage(1950-1970) it sought to achieve durable goods substitution, and in its third stage (1970-) capital goods substitution (Garza , 1984: p.12)

[25] The degree of success of that policy is reflected in the increasing share of the industrial sector in the DNP, which rose from 12.9% in 1930 to 22.5% in 1970 (Garza, 1984: p.13).

[26] (Gilbert & Ward, 1985: p. 88; Jury, 1983: p. 192-194).

[27] Import substitution induced an even greater degree of economic concentration in the MAMC in almost every sector of economic activity. With 17.8% of the population in 1970, its share in the GNP rose to 37%. In the commercial sector Mexico City accounted for 50% of the total value of national sales and of capital investment, and in the service sector for 55% and 58% of these two indicators; for 41% percent of all the jobs in higher education, 44% of all specialized medical positions, 69% of all government employees, 60% of all the bank deposits, and 90% of all bank offices (SPP, 1979).

[28] (Garza,1984: p. 3-24; Pommier, 1982:p. 352; Schteingart, 1983: p. 284).

[29] (Newel & Rubio, 1984: p. 107- 111; p. 146).

[30] Public spending increased from 23.6% of the DNP in 1970 to 36.6% in 1975. After a long historical period of frugality and discipline in foreign lending and monetary management, there began a period of rapidly growing foreign debt and inflation exacerbated in the years of the short-lived oil boom with its unrealized hope of long-term huge returns. (Newel & Rubio, 1984: p. 124-146; Gentelman, 1984: p. 34-60).

[31] (Scott, 1983: p. 18-21, 239-245; 248).

[32] (Scott, 1983: p. 18-21, 239-245; 248).

[33] (Scott, 1983: p. 244; Gilbert, 1985: p. 57).

[34] (Campbell & Wilk, 1986: p. 305-306, 312-313 notes 22, 23; Jury, 1983: p. 195; Schteingart, 1983: p. 292).

[35] (Garza, 1981; Scott, 1983: p. 197, 243; Newel & Rubio, 1984: p. 109; Schteingart, 1983: p. 284, 290).

[36] (Jury, 1983: p. 192).

[37] (Brennan, 1982; Campbell & Wilk, 1986: p. 299, 312).

[38] (Jury, 1983: p. 192-193).

[39] (Campbell & Wilk, 1986: p. 307-309; Ward, 1986: p. 222-225).

[40] (Scott, 1983: p. 237-235).

[41] Mexico City shows locational advantages in the production of durable consumer goods and capital goods, and disadvantages in the production of intermediate goods, (Garza, 1984: p. 7-20).

[42] (Stern, 1984; Garza, 1982; Vining, 1985, 1986).

[43] INEGI, XI Censo de Población y Vivienda 1990: Cuadro 4.2.

[44] (Scott, 1983: p. 266).

[45] (Scott, 1983: p. 250).

[46] (Bataillon, 1968: p.31).

[47] (Scott, 1983:p. 107-109).

[48] (Gilbert, 1985: p. 88).

[49] From 1950 to 1960 they registered annual growth rates of 13.7%, 11.1% and 10.3% respectively.

[50] (Unikel, 1971: 250, 255, 269; Schteingart, 1981: 172).

[51] (CCCR, Programa de Reservas Territoriales, 1986).

[52] A variant of the *vecindad* appeared in the forties. Vacant lot owners in higher income areas rented them to poor tenants to erect provisional shacks. These benefited from urban services and easy access to employment areas. Owners would invariably end up trying to evict the tenants as land value would rise. (Cornelius, 1975: p. 26; 1976: p. 252; Turner, 1971 p.1-4; Ward, 1981: p. 39; 1976: p. 252; Lomnitz, 1977: p.19;).

[53] (Diario Oficial, December 31, 1941).

[54] (Gilbert, 1985: p. 87).

[55] Around 2 million residents held on to them, passing them on to heirs, due to their prime location and virtually rent-free occupation in spite of their ruinous condition (Turner, 1971: p. 1-4; Cornelius W. 1975: p. 26-28; Lomnitz, 1977: p. 19; Ward, 1981: p. 39; Schteingart, 1988: p. 273- 274).

[56] (Stephens, 1983: p. 8-9).

[57] Some irregular settlements in the DF can be traced to the twenties and thirties, but it was during the forties that they began to grow rapidly. (SMP, 1982: p. 162; Gilbert & Ward, 1985: p. 72- 75; Turner J., 1971)

[58] (Sociedad Mexicana de Planificación, 1982: p. 162).

[59] (Perló M., 1979; González Navarro M., 1974; Furtak R. K. 1974).

[60] His administration expropriated 48.5 hectares of land for regularization and distribution to 28,272 families (Memoria del DDF 1940-46).

[61] The literature has linked their involvement and support of the settlers with their failed attempts to win the nomination of the party to the presidency.

[62] (Gilbert & Ward, 1985 q. Cisneros, n.d.).

[63] Uruchurtu's policy against the irregular settlements was used as a political weapon by the government party's congressional block to oust him from office in 1966, when he

evicted by force the irregular settlers of Santa Ursula Coapa (Lugo Medina & Bejarano Gonzalez, 1981:p. 245).

64 (Sociedad Mexicana de Planificación, 1982: p. 162).

65 (Cornelius, 1975: p. 185).

66 (DGDUV-SM, 1982; Iracheta A., 1983).

67 (Cornelius, 1975: p. 229).

68 DO, March 28,1941.

69 Azuela de la Cueva, 1989: p.99.

70 (Perlo, 1979: p. 811; Gilbert, 1985: p. 157, 192).

71 Cornelius, 1975: p. 160; Gilbert & Ward, 1985: p. 189. Quoting Cornelius: The informal and rather diffuse nature of patron-client networks acts to control the allocation of the limited resources available to the poor. Indeed it is through this system that the government manages to curtail demands for expensive, generalized benefits that might "overload" the political system.

72 (Gilbert & Ward, 1985: p. 95; p. 251; Cornelius, 1975: p. 26-27).

73 This classification based on origin was suggested by Gilbert and Ward. Earlier classifications distinguished 3 loosely defined types based inconsistently on physical form and origin: *ciudades perdidas*, *colonias paracaidistas*, and *colonias proletarias*. (Bataillion, 1968; Cornelius, 1975; Stephens, 1981).

74 (Cornelius, 1975: p. 30-35).

75 (Gilbert & ward, 1986:p. 92).

76 (Azuela de la Cueva, Tlaxcala Conference Acts, 1981; p. 218).

77 (Cornelius, 1975: p. 30-35). *Colonia proletaria* residents display a better economic situation than those of the other types. In the Ajusco case study, the average family income was twice the minimal wage (López Rivera C., 1978).

78 Irregular company-sponsored subdivisions were created within the DF after the adoption of the 1941 regulations and the 1953 ban. The explanations offered refer to the existence of earlier authorizations carried out after these dates or to pseudo-invasions, rather than to government connivance as in the SM (Gilbert & Ward, 1986: p. 88).

79 (Guerrero et, al. 1974; Schteingart, 1978, 1983; Gilbert & Ward, 1986).

80 (Diario Oficial, March 25, 1954).

81 (Ward, 1981).

82 (Diario Oficial, May 25, 1954).

83 Alatorre y Camargo, 1986; (Azuela de la Cueva, 1989:p. 120).

84 Many settlers paid their way into these settlements by working in their weekends on the primitive urbanization works undertaken by the *comisariados ejidales*.

85 (Diario Oficial, April 16, 1971).

86 (LFRA, Chap. III, art. 91; Diario Oficial, January 17, 1984).

87 The 1975 SM code included a penalty from six months to ten years jail and a fine of $M 500,000 for illegal subdividers.

88 DART brings together squatters, land owners, and developers to negotiate. If an agreement is reached, then the periodic payments from residents to the owners are defined, as well as the basic services that these have to provide. Otherwise, government expropriates, compensates the owners and resells to the residents.

89 (Diario Oficial, August 7, 1973; restructured DO November 8, 1974).

90 According to a CORETT report, of a total of 206.1 square kilometers expropriated nationally through CORETT, 191.5 square kilometers were located in the MAMC (Stephens R. 1981).

91 (Ward, ed. Pacione, 198: p. 52; Gilbert & Ward: p. 91, 1985).

92 DO, April 3, 1979.

93 (DO, May 20, 1976; Ward, ed. Pacione, 1981:p. 52).

94 (Ward, 1981:p. 52-53; Azuela de la Cueva, 1989:p. 124).

95 (Azuela de la Cueva, 1989:p. 124)

96 (Tlaxcala Conference, 1981. Gilbert & Ward, 1986).

97 Estadisticas sobre Vivienda, SEDUE. Catalogo CIHAC, 91:p. 374; Mexico.

98 In the 1983-90 period only 5.6% of the over a million finished public housing dwellings built were rentals (CIHAC 91:p. 370).

99 Ley de titulos y Operaciones de Credito (DO, Diciembre 31, 1946).

100 Ley sobre el Regimen de Propiedad y Condominio de los Edificios Divididos en Pisos. (DO, December 15, 1954; January 11, 1955).

101 The exception was the CINVA center of the OAS in Bogota which played a pioneering role in Latin America.

102 (Azuela dela Cueva, 1989)

103 CIHAC, El Mercado Mexicano de la Construcción, 1987/88, p. 135

104 INEGI, X y XI Censos de Población y Vivienda.

105 DO, April 24,1972.

106 INFONAVIT reached in 1990 a total affiliation of 672,289 firms employing 6,754,918 workers (1990). 33% of these affiliated workers and 22.2% of the factories are in the DF, while 8.7% and 5.9% respectively are in the SM.

107 (CIHAC 91:p. 366)

108 (Garza & Schteingart, 1978; Stephens, 1983: p. 23)

109 (Anuario Estadistico del DF, 1984: Cuadro 2.1.20)

110 Official report of the Comite de Reconstrucción del Area Metropolitana de la Ciudad de México (1986). México.

111 DO, May 9, 1986.

112 (CIHAC, 1987/88:p. 109)

113 (SEDUE: CIHAC 91)

114 (CIHAC 91:p. 363)

Chapter III

THE EJIDOS AS A FACTOR IN MAMC EXPANSION

Next one must achieve a greater understanding of how the *ejido*, the dominating tenure form on the MAMC periphery, has influenced in various ways the character of the city's growth, increasingly becoming a major obstacle to orderly planning. The role played by *ejido* property rights within the process of urban expansion on the rural periphery is examined, and the effects that are structurally related to these property rights are brought to light, particularly as a major contributing factor to irregular settlement formation. The increasing importance of *ejido* land in the expansion process of the MAMC and its dominant role as the site of irregular settlement activity is fully documented, and an explanation of the role played by *ejido* property rights within this process is suggested after alternative approaches in the literature are discussed.

EJIDO AND *COMUNAL* LAND PROPERTY RIGHTS: A LEGAL VIEW

Mexican cities in general and Mexico City in particular are surrounded by *ejido* and *comunal* land. These two forms of property are owned corporately by peasant communities. They are both non-marketable corporate property which cannot be alienated by normal means.[1] Article 52 of the Federal Law of the Agrarian Reform states clearly:

> The rights over agrarian lands held by the *ejido* communities are inalienable, inextinguishable, can not be mortgaged nor sold, and therefore, can not under any circumstance nor by any means be sold, ceded, transmitted, mortgaged, neither totally nor in part, for which reason any operations, acts or contracts which have been carried out, or are intended to be carried out in contravention of this law will be considered to be inexistent.

Although similar in almost every other respect to *ejido* property which was established by the constitutional convention of 1917 during the Mexican Revolution, *comunal* property rights have their origin in pre-Columbian and colonial times when the native villages were granted land by the Spanish crown. These property rights became the basis of restitution to those native peasant communities which had lost their land in the past. Much of that land was appropriated from the villages in the nineteenth century, but began to be returned to them by the restitution program initiated by the Mexican Revolution in 1915 as part of general land reform. *Ejido* land, by contrast, is not apportioned on the basis of ancient claims, but as a result of new rights established in article 27 of the 1917 Constitution of the revolutionary government of Venustiano Carranza which allowed peasants to petition the government for land endowments. An agrarian reform program emerged in the years that followed which divided the large agricultural estates and granted land to many peasant groups or communities who never owned any, or who were in need of additional land for agricultural purposes. Endowment of *ejido* land became by far the dominant form of the two in the agrarian reform program instituted by the Mexican post Revolutionary regimes.

Another difference is that *comunal* land is owned by the total village, which includes all the residents engaged in agriculture with more than five years of residence there, while *ejido* land belongs only to the officially recognized members of the *ejido* community, which very often may not include the total village population.[2]

Besides *ejido* and *comunal* property two other forms of land tenure can be identified on the periphery of Mexican cities: private and national property. Private property of land in the rural areas is subject to a upper size limit since 1917. The large estates were outlawed and only *pequeña propiedad*, small properties, within the upper limit of 100 hectares for irrigated agricultural land and 200 hectares for seasonal land, are legally allowed. Private property rights are only granted by the 1917 Constitution on the surface of the land which can be sold, transfered or mortgaged, unlike the subsurface property rights which can only be granted in concession or leased by the state to private interests. Surface property rights of national lands can be alienated by the government and turned into private property by market and non-market means when required by the public interest.

As long as *ejido* and *comunal* property maintain their original status, there is no legal way to use the land, much less to sell it, for urban development purposes. To change its status, it must go through a complex and lengthy expropriation process which requires nothing less than a presidential decree on grounds of higher public interest. It also requires that the government pay compensation to the expropriated *ejido*. Compensation to be paid to the *ejido* is assessed on the agricultural value of the land as distinguished from the much higher commercial value. The use of this money is highly restricted in the agrarian legislation and does not normally go directly to the pockets of the *ejidatarios,* but to the *ejido's* account with FIFONAFE, the national economic development fund of the *ejidos*.[3]

The grounds for terminating *ejido* property are specifically listed in the law as proven public interest purposes, which must be clearly superior in their social significance to the present use of the land. The public interest purposes which may justify an expropriatory action by the state listed in the agrarian reform legislation are the following: the establishment, operation or conservation of a public utility; the opening or widening of railroad lines, highways, streets, bridges, etc.; and the creation of an enterprise of unquestionable benefit to the community.[4]

Urban development in general, without specifying a particular need, was not included among the public interest purposes justifying *ejido* expropriation until 1984.[5] An amendment was then made to the law: "The following are public interest causes:...I-VI. The establishment, improvement, conservation and growth of the population centers whose ordering and regulation is foreseen in the national, state and municipal plans of urban development and housing." The establishment of land reserves for urban expansion is implicitly included in this clause as a necessary part of urban development.[6]

Part of an *ejido* can also be privatized for its own members through the highly restricted and equally complex procedure of creating a *zona urbano ejidal* (ZUE), a village for the *ejido* members. The *ejidatarios* then become the private owners of the residential lots assigned to them. Since the *ejidatarios* themselves are free recipients of the ZUE parcels, no compensation payments are involved. There also existed a third route to achieve privatization: land barter between *ejidos* and private landowners, also requiring presidential approval; this option was included in the Agrarian Code in 1942 and later outlawed in 1971 by the Echeverria government.

Confiscation of property without compensation is outlawed in the Mexican Constitution. As a direct consequence, government has a legal obligation to pay compensation to any *ejido* that is subject to expropriation. In the case of the establishment of a *zona urbano ejidal* no compensation payments to the *ejido* are granted since the *ejido* members themselves are the free recipients of the ZUE parcels. Compensation, however, in all other cases, is assessed according to the agrarian legislation, on the basis of the agricultural value of the land, regardless of its commercial value (Ley Federal de Reforma Agraria, art. 122-I, II, III). This principle has been uniformly applied to *ejido* expropriations.

Within the basic principle that compensation is assessed on the agricultural value of the land some variations depending on the particular purpose of the expropriation were introduced in 1984.[7] Where the expropriation is for public works—or as added in 1984—for the establishment of urban land reserves or social interest housing programs, the compensation amount is assessed on the agricultural value of the land and should enable the *ejidatarios* to acquire an area of land for agricultural purposes of the same agricultural quality and size anywhere else, without specifying where.[8] The *ejidatarios* in the urban periphery can thus be offered agricultural land anywhere in the country as compensation. If two thirds of the *ejidatarios* cast their votes to reject reinvestment in agriculture, the law offers them the option of receiving the same equivalent amount of money as a deposit in FONAFE, the official agrarian investment fund, for other investment programs approved by the Ministry of the Agrarian Reform.[9]

Where the purpose of expropriation is "the establishment, improvement, conservation or growth of population centers whose regulation and ordering is foreseen in the national, state, and municipal plans of urban development and housing" compensation is assessed on the agricultural value of the land, plus a 20% share in the net profits of the land developed, and two average urban size lots in the developed area, for each *ejidatario*. This 1984 addition, which can hardly be distinguished from the one mentioned in the previous paragraph concerning urban land reserves, introduces a large measure of ambiguity.[10]

The Agrarian Reform Law also establishes in its 1984 additions that where the expropriation of an *ejido* is done for the purpose of regularizing an existing irregular settlement, the compensation is assessed on twice the agricultural value of the land, plus a 20% share in the net profits of the regularization program, and two average urban size lots in the regularized settlement, for each *ejidatario*.[11]

EXPANSION OF THE MAMC ON *EJIDO* LAND

Expansion of Mexican cities in general, and Mexico City in particular on its surrounding rural periphery, is taking place increasingly on *ejido* and *comunal* land. In spite of the seemingly great difficulties of converting *ejido* and *comunal* property to private status, much of the expansion of the MAMC has somehow occurred on this land. Between 1940 and 1975, 48% of the total growth of the MAMC took place on *ejido* land (table 9). This figure, however, masks the important fact that the proportion of *ejido* land being urbanized has been increasing over time. Urbanization of *ejido* land expanded from minimal levels in the forties to total dominance of the urbanization process in later decades. The trend is most visible in the DF data, which are available for intermediate points in time (table 9).

In the early part of the 1940-50 decade, *ejido* land was sparingly converted to urban uses in the MAMC. Limited amounts of land were expropriated for industrial purposes such as the creation of the Vallejo industrial zone and the PEMEX refinery in Azcapotzalco. Since 1946, however, the government has greatly expanded its use of *ejido* land for a wider variety of purposes, beginning with the creation of the UNAM campus. Taken as a whole, during the 1940-50 decade only 12.6% of total urban development had taken place on *ejido* land in the DF. In the 1950-60 decade, however, that percentage increased to 41.6%, and between 1960 and 1975 rose further to around 66%, as illegal use of *ejido* land for low income irregular settlements began to grow rapidly (table 9). In the State of Mexico MAMC—the most active front of recent urban expansion—that same trend is visible. Of all the land converted in the SM to urban settlements in the 1940-75 period taken as a whole, 49.4% was taken from the *ejidos*. This proportion rose to 61.5% in the 1975-82 period (table 9).

Did this growing use of *ejido* and *comunal* land—in absolute and relative terms—occur within the framework of the strict requirements and procedures of the agrarian legal system? This is a very important question that not only has moral, political, and economic implications, but also has the greatest significance from a physical planning standpoint. Initial illegality determines a chain of physical negative impacts which run counter to the very essence of planned development.

High social costs are imposed by illegal conversion of *ejido* land. Even where illegality does not lead to irregular settlement activity, corruption of the entire agrarian land system is encouraged as has been well documented in the literature. The case studies, which will follow at the end of this chapter, document the corruption and serious social conflicts between *ejidatarios*, settlers, government and other interested parties. This illegal process drains away social energy, clouds the planning process and distorts property rights.

Investigations have shown that only part of *ejido* land has entered the urban system through the legal channel of government expropriation, land barter (before 1971), or legitimate establishment of a *zona urbano ejidal* . An important proportion of *ejido* land has been urbanized illegally without being freed from *ejido* status. Even where the land was legally disestablished by expropriation, there very often exists a covert violation of the agrarian law. *Ejido*

Table 9. Expansion of the MAMC (sq.km.) and Share of *Ejido* Land in its Growth Area (1940-1982)

Years	DF		SM		MAMC	
	All Types (sq. km.)	*Ejido* (%)	All Types (sq. km.)	*Ejido* (%)	All Types (sq. km.)	*Ejido* (%)
Land Use in 1940[ß]						
Urbanized Area	117.5	-	0.0	-	117.5	-
Rural Area	1381.5	75.4	936.5	63.0	2318.0	70.4
Total Region	1499.0	-	936.5	-	2435.5	-
Land Use in 1975						
Urbanized Area	540.7	-	251.3	-	792.0	-
Rural Area	958.3	87.8	685.2	67.9	1643.5	79.5
Total Region	1499.0	-	936.5	-	2435.5	-
Land Use in 1982[ß]						
Urbanized Area	541.4	-	489.9[«]	-	1031.3	-
Rural Area	957.6	87.8	1238.1	51.6‡	2195.7	67.4
Total Region	1499.0	-	1728.0	-	3227.0	-
Urban Area Increases (1940-82):						
1940-50	111.8	12.6	0.0	0.0	111.8	12.6
1950-60	83.9	41.6	102.8[«]	n.a.	186.7	n.a.
1960-70	147.5	66.6	92.2[«]	n.a.	239.9	n.a.
1970-75	80.0	65.2	56.3[«]	n.a.	136.3	n.a.
1975-82	0.7	n.a.	238.6[«]	61.5	239.3	61.3
Total 1940-75	423.2	47.2	251.3	49.4	674.5	48.0
Total 1940-82	423.9	47.2	489.9[«]	55.3	913.8	51.4

Sources: Schteingart (1982:p. 286, tables 1,2); [«] Iracheta (1984:p. 58, t. 4, 5). (CCCP, 1986:p. 39, table XI.). [ß]11 SM municipalities are included in the 1940 and 1975 estimates, 15 in 1982.

land is often expropriated by the government for purposes recognized in the law, but the land is soon diverted to other uses. Expropriation in these cases only serves as a cover leading to the diversion of the expropriated land to illegal uses, casting a shadow of doubt on the legitimacy of the original expropriatory act.

Ejido land can be classified from this standpoint into four distinct categories: land legally converted and urbanized for its intended purposes; land legally converted and illegally diverted from its intended purposes; land illegally converted and given fictitious legal cover; land illegally converted and irregularly settled.

Within the first category, land legally converted and used for the intended purpose, one can find that *ejido* land has been used for public utilities and services, highways, industries, university campuses, public housing units, public parks, sport installations and airports. The National university (UNAM) and the National Polytechnic Institute (IPN) campuses were both established on *ejido* land. An area of 747 ha. of *ejido* land were expropriated mostly from the Tlalpan *ejido* for UNAM, and 359 ha. from the San Pedro Zacatenco *ejido* for the IPN campus and for Pemex, the state oil company.[12]

Expropriation by presidential decree has provided government agencies with large tracts of land for public housing purposes since the fifties, as the government became involved in public housing. The housing units of San Juán de Aragón, Unidad Atzacoalco, Unidad Lindavista Vallejo and Cuautitlán Izcalli, to mention just a few of the largest public subdivisions, were developed on *ejido* land. AURIS and ODEM, the urban development agencies of the State of Mexico, have been instrumental in the expropriation of *ejido* land for their social interest housing programs.

A second category of *ejido* land was legally converted but diverted for illegal purposes. Schteingart and García Peralta have documented that much *ejido* land in the MAMC was actually diverted for private commercial purposes very different from those originally sanctioned.[13] These authors found that much *ejido* or *comunal* land expropriated for the public agencies and intended for low income public housing was sold by the agencies to private interests at minimal, below market prices, for commercial purposes not in agreement with the original expropriatory decree. *Ejido* land obtained by AURIS at minimal cost for housing programs was subsequently sold at lower than market prices to private interests for commercial developments to solve its financial crisis. ODEM, an official body of the State of Mexico, sold great parts of developed *ejido* land expropriated for Cuautitlan Izcalli to private construction and real estate companies, at lower than market prices, for higher income housing.[14]

These authors also include within this category *ejido* land that was legally bartered by private interests for upper-class developments, before this form of privatization was excluded from the agrarian law in 1971 by the Echeverria administration. Elite developments such as Jardines del Pedregal and San Miguel Tecamachalco, and middle class developments such as Ciudad Satélite were established on bartered *ejido* land. They estimated that 41% of urbanized *ejido* land in the SM-MAMC has been used for high and middle income residential developments. These authors include these exchanges in the category of illegally diverted land on grounds that use of *ejido* land for upper-class housing is an illegitimate use of *ejido* land, even if it was legally carried out.

The third category, *ejido* land illegally converted and given fictitious legal cover, is understandably very difficult to detect. Ward investigated the subject nationally with a team from the Ministry of Human Settlements and found, for example, that various developers acquired illegally large areas of *ejido* land in the second-home belt of Morelos. They paid several years' land taxes to obtain official receipts, and used them to notarize and register their ownership in the official land registry. Then they subdivided and sold parcels as upper class residential developments.[15]

Another means used to disestablish *ejido* land under a false legal cover is appropriation of *ejido* land by neighboring private claimants of property, often without any legitimate basis. However these illegal practices worked, they account for the important numerical discrepancies between the amount of legally privatized *ejido* land and the total *ejido* acreage used for higher income developments in the State of Mexico. These differences are cited as evidence for the existence of land within this category.[16]

The fourth category, *ejido* land which is occupied without going through the legal disestablishment procedure and is urbanized as irregular settlements, has become the most extensive of all. Occupation of the land has taken place in a variety of ways: through invasion by squatters, illegal sale and development by companies, and under the false cover of the ZUE fiction.

When *ejido* land enters the urban system without being legally disestablished as in this case, the whole future trajectory of that land is distorted. A vicious circle of irregular development begins: holders of the land are not able to establish legal title to the property, and as a consequence can not obtain official permits for subdivision and sale; the subdivision is not incorporated into the city's service networks and tax base, and is excluded from any public investments; buyers of lots are not able to obtain legally valid title deeds, nor credits or mortgages. Illegal conversion of *ejido* land thus leads to the formation of irregular settlements. From a wider planning perspective illegally disestablished *ejido* land escapes the basic controls of the planning system, which can only work within the framework of the legal system. The basic parameters of where, when, and how to develop land is decided solely by the illegal *ejido* subdividers and settlers in a chaotic manner and in ways inappropriate for urban development.

IRREGULAR SETTLEMENTS ON *EJIDO* LAND

Although not all the irregular settlements are on *ejido* land nor is the totality of *ejido* land occupied by irregular settlements, it is nevertheless true that taken as a whole, a) most of the irregular settlements are located on *ejido* land, and b) a much higher proportion of *ejido* land is occupied by irregular settlements than is the case with non-*ejido* land.

In the DF-MAMC area, according to the DF Development Plan (PDUDF), 90% of the irregular settlements are on *ejido* and *comunal* lands, with only 10% being located on private lands.[17] This fact is reflected in the land regularization programs of the irregular settlements conducted by various government agencies. These regularization programs have almost become synonymous with *ejido* regularization.

In the State of Mexico metropolitan area (SM-MAMC), the most active growth front of the metropolis, 78.9% of the land area occupied by the irregular settlements is *ejido* land although *ejido* land constitutes around 55% of the total stock of urbanized land (1.43 times its expected share) (table 10). The opposite is true for non-*ejido* land on which only 21.1% of the area of the irregular settlements is located, although 45% of the land in the urbanized SM-MAMC belongs within this category (0.47 of its expected share). A strong association between *ejido* land and irregular settlements is likewise reflected in the way the land incorporated to the SM-MAMC is used (see table 11): while only 13% of the non-*ejido* lands in the SM metropolitan area are occupied by irregular settlements, in *ejido* lands that proportion rises to 40.3%, three times the former proportion.

The use of *ejido* land for irregular settlements has increased over time. Some of the government policies reviewed have reinforced this tendency on both the demand and supply sides. The end of land invasions and company sponsored irregular settlement activity, which also occurred dominantly on *ejido* land, left the ZUE settlements as the only source of cheap land that is satisfying settler demand in its totality. The regularization program and the reserve program have, on the other hand, induced more illegal sales of *ejido* land for irregular settlements than ever. The net result has been that the near totality of irregular settlement activity has been concentrated on *ejido* land in the periphery.[18]

Table 10. Land Area of the SM-MAMC (sq. km.) Classified by Type of Settlement, Land Use and Tenure (1940-1982)

Type of Settlement	Land Tenure *Ejido*	Non-*Ejido*	Total
a) Regular Settlements	122.3 (43.8%)	156.7 (55.5%)	279.0 (100.0%)
b) Irregular Settlements	109.2 (78.9%)	29.2 (21.1%)	138.4 (100.0%)
c) Residential Use Subtotal (c=a+b)	231.5 (55.5%)	185.9 (44.5%)	417.4 (100.0%)
d) Non-Residential Use Industry and Services	39.4 (54.9%)	32.4 (45.1%)	71.8 (100.0%)
e) Total (e=c+d)	270.9 (55.3%)	219.0 (44.7%)	489.9 (100.0%)

Source: Iracheta, 1984: p. 59, table 5; p. 292, table 7

Table11. Land Area of the SM-MAMC (sq. km.) Classified by Type of Settlement, Land Use and Tenure (1940-1982)

Type of Settlement	Land Tenure *Ejido*	Non-*Ejido*	Total
a) Regular Settlements	122.3 (45.2%)	156.7 (71.8%)	279.0 (57.0%)
b) Irregular Settlements	109.2 (40.3%)	29.2 (13.3%)	138.4 (28.2%)
c) Residential Use Subtotal (c=a+b)	231.5 (85.5%)	185.9 (85.2%)	417.4 (85.2%)
d) Non-Residential Use Industry and Services	39.4 (14.5%)	32.4 (14.8%)	71.8 (14.8%)
e) Total (e=c+d)	270.9 (100.0%)	219.0 (100.0%)	489.9 (100.0%)

Source: Iracheta, 1984: p. 59, table 5; p. 292, table 7

Table 12. Land Area of the SM-MAMC (sq. km.) Classified by Tenure and Type of Settlement (1940-1982)

Type of Settlement	Land Tenure *Ejido*	Non-*Ejido*	Total
Regular Settlements	n.a.	n.a.	2,599,049 (36.8%)
Irregular Settlements	3,54,688 (5.0%)	923,745 (13.2%)	4,448,433 (63.2%)
Total	n.a.	n.a.	7,047,482 (100.0%)

Source: Iracheta, 1984: p. 59, table 5; p. 292, table 7

EXPLANATIONS FOR *EJIDO* DOMINANCE IN URBAN DEVELOPMENT AND IRREGULAR SETTLEMENTS

Important questions can be raised at this point: why has irregular settlement activity been taking place dominantly on *ejido* land? Furthermore, why have the *ejidos* found it attractive to supply this secondary market with its low prices and high risks? The strong association of the irregular settlements with *ejido* land can not be treated as merely accidental, and requires a rational explanation. Neither can one ignore another related more general question: why has *ejido* land increasingly been found more attractive than private land by government for all types of development, legal and quasi-legal, when expropriation or regularization of *ejido* land is far more complex and protracted than acquisition of private land.

As pointed out in the literature and explained in the preceding chapter, increased illegal urbanization of *ejido* land in the MAMC since the fifties is linked to the implementation of a series of important government land control policies. These policies triggered intensive illegal subdivision and sale of *ejido* land for irregular settlements. The specific policies referred to are: the ban on new subdivisions in the DF by mayor Uruchurtu from 1953 to 1966; the government clamp-down on illegal company-sponsored subdivisions after the residents' payments strike of the early seventies; the firm stand taken against squatting since 1977, which has brought an end to land invasions in the MAMC; the regularization programs of CORETT and FIDEURBE to normalize the land tenure situation in the irregular settlements since 1973 and finally the attempt of the Ministry of Human Settlements to create land reserves for urban expansion based mainly on *ejido* land.

The literature has sought explanations for the effects that these policies had on illegal *ejido* sales and irregular settlements activity. Some authors have come up with explanations which merit close scrutiny. Authors Gilbert and Ward have suggested four principal reasons which can be summarized as follows:[19]

First, the political autonomy of the *ejido* made it possible for the *ejidatarios* to act on their own and evade the limitations and restrictions of local government on irregular settlement formation, turning the *ejidos* into an effective escape valve for these activities.

Second, much of the *ejido* lands surrounding the Mexican cities and the MAMC have great strategic and economic urban value inducing the *ejidatarios* to commercialize that land illegally, particularly where its agricultural worth is minimal.

Third, regularization of the irregular settlements on *ejido* land acts as a powerful economic incentive to the *ejidatarios* to continue selling their land illegally in the informal market to settlers in order to benefit twice: first from initial sale price and then from subsequent compensation payments.

Finally, illegal settlement on *ejidal* land has flourished because of the political patronage that *ejidatarios* enjoy. The *ejidatarios* manipulate and benefit from a special situation based on privilege, which allows them to sell *ejido* land in connivance with officials in the Agrarian Reform Ministry.

However, these explanations are incomplete and taken as a whole leave out the most significant aspects of the problem. Inadequate conclusions are drawn as a result. The flawed perception that *ejido* behavior is the result of a situation of privilege is uncritically accepted in these explanations. The argument is put forward instead in this study that it is far nearer to the truth to view the *ejidatarios* as victims than as offenders. Blaming the victim distorts the substantive issues of *ejido* illegality. A close review of these explanations as they relate to the policies previously referred to is necessary before suggesting an alternative approach.

In reference to the first of these policies—the DF development ban—the question which is asked in the literature is how any development at all took place during the Uruchurtu period (1953-66), when strict enforcement of the ban was seriously attempted by the mayor. Gilbert and Ward use the "escape valve" notion just mentioned to explain that effect:[20]

> Given the high level of demand for self-built lots in the fifties and sixties and the strict controls in force in the DF [against land development since 1953], there was pressure on the *ejido* lands on which mayor Uruchurtu's department held little authority. Formally he could only block service applications for service provision.

In other words, the ban redirected part of the land demand to the *ejidos* of the DF, which acted as an escape valve for urban growth pressures. The autonomy of the agrarian system, which is not controlled by the local authorities and holds important land resources in the DF, made it possible for the *ejidos* to overcome the ban.

One must agree that the *ejidos,* which underlie the agrarian laws and authorities, function as autonomous enclaves in the urban areas. They are, for all practical purposes, largely independent of the city authorities. This protected status, which is rooted in *ejido* history, legislation, and power politics, made it possible for them to evade the controls of the city government. Thus, *ejidos* engaged in land subdivision and sales under the cover of the ZUE legislation, in spite of the efforts of the most strong-handed mayor the DF ever had to implement the ban in the DF area from 1953 to 1966.

Mayor Uruchurtu did deny development permits and services. His authority, however, stopped at the doorstep of the *ejidos* as it did at the threshold of the State of Mexico. Both the *ejidos* and the State of Mexico played the role of escape valves to the controls he was trying to impose. This combination of factors resulted in irregular settlements on *ejido* land in the DF and increased settlement activity in the State of Mexico. Instead of producing a more adequate urban form as intended by the non-growth policies of mayor Uruchurtu, the MAMC took on an increasingly irregular and chaotic character.

The same escape-valve rationale is used by these authors to explain the very similar effects observed when the government adopted a policy of clamping down on the company-sponsored irregular subdivisions after the settlers' general payment strike against the companies in the early seventies. Invasions on *ejido* land and sale of *ejido* land in the ZUE's greatly increased as a consequence of the clamp down, which closed an important source of land for low income people.

The escape-valve rationale is also applied successfully by Gilbert and Ward to explain the effects of the López Portillo government's (1977-82) policy of bringing an end to forty years of government-tolerated or sponsored invasion of land by irregular settlers, initiated in Mexico City by mayor Javier Rojo Gómez (1940-46) in 1940. The final elimination of the invasion option brought about increased pressure on the only remaining alternative open for the irregular settlers, that of buying themselves into *ejido* subdivisions operating under the ZUE cover. In the authors own words:[21]

> Recently also, firm action has been taken to eradicate new invasions as they occur in the DF, which, while effective insofar as land control is concerned, is having two principal effects: The first is to increase densities in existing areas, the second to increase pressure for land sales in *ejido* lands, which, as we have seen, form a relatively "protected" area.
> Recently, the exercise of tighter controls against invasion and the clamp-down on company-sponsored subdivisions have left the *ejidos* as the only real source of cheap land.

However, when it comes to explaining the increase in illegal subdivision and sale of *ejido* land following the regularization programs implemented by CORETT, Gilbert and Ward resort to the explanation that the CORETT compensation payments acted as an incentive to the *ejidatarios* to increase legal sales in order to benefit doubly: first, selling illegally to irregular settlers and later obtaining additional compensation through regularization. The sales frenzy was brought about when the *ejidatarios* learned that the CORETT charter empowered that institution to expropriate still unoccupied *ejido* land for the purpose of creating new commercial housing subdivisions to generate profits for FONAFE. The *ejidatarios*, according to this view, obviously preferred sale and regularization with its double benefits over simple expropriation, and entered into a frenzied race of illegal sales:

> The requirement that ejidatarios receive compensation for the 'loss' of their land has quickened the pace of illegal *ejido* sales. It made sense for ejidatarios to sell their plots, particularly after the establishment of CORETT whose initial charter empowered the agency to acquire vacant *ejidal* land and to commercialize it for housing purposes.[22]

Seemingly to address the question of the failure of the Ministry of the Agrarian Reform to control the illegal sales of *ejido* land, Gilbert and Ward introduce their explanation of the political influence of the *ejidatarios* and the interests that guide the Ministry's policy:[23]

> The sale of *ejido* land is overlooked by the Agrarian Reform Ministry, because of the relative political weight of the agricultural workers' sector (the CNC)—one of the three pillars of the ruling PRI. Support from this mass base is critical to the political influence that the ministry holds, as well as to the stability of the overall regime.
> There are also other reasons. Illegal land sales are lucrative not only for *ejidatarios* but also for ministry officials. Engineers who draw up and authorize the certificates issued to low-income buyers take their cut part of which is handed on to their bosses.

The notion that the *ejidatarios* seek to profit from regularization is also used by Ward to explain the greatly increased illegal sales of *ejido* land that the intended policy of establishing urban growth reserves by the Ministry of Human Settlements triggered a few years later. The *ejidatarios*, according to this view, once again preferred illegal sale and regularization over expropriation, to increase their benefits:[24]

> The restructuring of the public administration (1977) and the allocation of responsibility for human settlements to a ministry of the same name has given weight to recent attempts to create land reserves often at the expense of *ejido* lands. This [establishment of the reserve] ironically, further promotes ejidatarios to subdivide and sell willy nilly, in order to guarantee an *ex gratia* initial payment and to elevate the value of the second payment that they can expect to receive via indemnization. The attempts to thwart the illegal sales or to quicken the takeover of *ejido* lands encountered resistance from the Agrarian Reform Ministry, and the suggestion of incorporating sanctions against *ejidatarios* were not even countenanced.

Ward finally comes to the conclusion on the basis of his previous premises:[25]

> Therefore the problem as articulated here is one of finding not so much a technical solution but a way through the political interests and rivalries embodied within Mexican public administration.

These explanations and conclusions, do not take into consideration the fundamental motives underlying *ejido* behavior. While it is true that the regularization program and the intended reserve program brought in their wake a large increase in *ejido* sales, it is an oversimplification, to say the least, to trace it to the CORETT compensations, which played at best a secondary role. Such an explanation is flawed, to begin with, by confusing the minimal amounts paid out as compensation to the *ejidatarios,* with the far greater proceeds of the regularization operation as a whole, which were designed to mainly benefit CORETT and

FONAFE.

As far as the expropriated *ejidos* are concerned—as distinguished from the agrarian establishment as a whole—they benefitted minimally from that compensation. Of the total amount of money collected in every regularization program, 10% was retained by CORETT to pay administrative costs, 67% went to FONAFE (now FIFONAFE) to be used nationally for **other** *ejidos* anywhere in the country, and only 23% entered the regularized *ejido's* account at FIFONAFE as compensation.[26]

Every regularization program initiated by CORETT or FIDEURBE since 1973 brought about a hostile reaction from the *ejidatarios* against these agencies.[27] Each of these programs produced mass mobilization of the *ejido* communities and organizations in the MAMC which joined the affected *ejidos* in solidarity to protest and defend *ejido* rights against the regularization agencies. It should have become obvious that more fundamental reasons than benefitting from CORETT compensation stood behind *ejido* opposition and behavior.

To view the *ejidatarios* facing expropriation as profiteering from a position of political privilege is deeply misleading. Gilbert and Ward have fallen into the trap of those interests that covet *ejido* land and would like us to believe that the *ejidatarios* are the profiteers that stand in the way of the public good. They have failed to take into account several crucial factors which should not only change the perception of *ejido* behavior, but also throw light on the government's increasing preference for expropriating *ejido* land over buying or expropriating private land. Several issues other are involved than those recognized by Gilbert and Ward and other authors such as Varley (1985), must be examined in order to gain a comprehensive view of *ejido* land dynamics in the urban peripheries.

A fundamental factor to be taken into consideration concerns the process of expropriation of *ejido* land and its varying significance for the *ejidatarios* and other interested parties. Land for the *ejidatarios* is not a simple commodity for speculation, but the source from which they draw a living. For the *ejidatarios* expropriation signifies the loss of their livelihood and their way of life. When they lose that livelihood and expropriation is carried out, they find that compensation payments are minimal and uncertain. The expropriation clauses of the agrarian legislation which prescribe that compensation should be assessed on the agricultural value of the land, even where double that amount is paid, keep the initial cost of expropriating *ejido* land far below market value. This fact alone—although there are other no less important reasons—should be enough to explain why the *ejidos* have tried by all possible means to circumvent government expropriation by illegally subdividing and selling their land beforehand, and why *ejido* land became an increasingly attractive target for expropriation by government and government-linked political interests.

Commercial land values in the expanding periphery of a growing city rapidly rise far above agricultural value.[28] In rural areas, agricultural and commercial value of land may be indistinguishable. However, in the periphery of an expanding city, paying for land on the basis of agricultural value can turn out to be the near equivalent of confiscation, considering the differential between commercial and agricultural value of land.[29] In the Miguél Hidalgo case study later surveyed in this chapter, commercial land values were on the average 125 times higher than the compensation assessed on agricultural value when regularization was undertaken in 1976. The differences are very often much greater.

The minimal compensation payments received by the *ejidatario* are well documented. The payments are recorded in the expropriation decrees published in the Diario Oficial. Going back to the older expropriations of *ejido* land in the DF, one learns that the total monetary compensation paid to the *ejidatarios* for 733 hectares in the *delegaciones* of Tlalpan and Contreras for the establishment of the UNAM campus, amounted to $95,270 MP ($19,643 US).[30] This is the equivalent of less than one Mexican cent for each square meter! Compensation payments have obviously risen over time; nevertheless, according to Schteingart the price paid to the peasants in the more recent decades stood below an upper limit of $11 pesos/sq.m. (0.08 US$/square foot in 1970 fixed prices) in 93.8% of all the land expropriated. The payments per *ejidatario* family are normally the equivalent of a few hundred dollars, seldom going beyond two thousand dollars.

The minimal cost of expropriating *ejido* land makes this a far more attractive option than buying private land for government projects. Government agencies increasingly cast their eyes on the land of the *ejidos*, which can be obtained at bargain prices, to solve their justified and unjustified needs. The greater the margin between the agricultural price of land and commercial price due to urbanization pressure, the greater the attraction that *ejido* land would exert on all the interests that are powerful enough to use government expropriatory powers to their benefit. Soon after *ejido* land is expropriated, some of it finds its way to the market, benefiting the political middlemen. The rightful owners, the *ejidatarios,* gain little, and become urban proletarians or petty merchants to survive. The artificially low prices of expropriated *ejido* land stimulates takeover and diversion of *ejido* land from its stated expropriatory purpose for all types of commercial uses not compatible with the special higher public interests restrictions of *ejido* land.

Protracted and almost endless administrative procedures lasting many years are necessary to actually obtain compensation payments in a time of rampant inflation; therefore the *ejidatarios* seek other ways to protect themselves. Many years transpire before some form of compensation actually reaches the *ejidatarios*.[31] The newspapers often report on unpaid compensation claims; one such report indicates that the *ejidatarios* of Cuemanco are demanding, after 24 years, compensation for the 36 hectares of land that where taken from them to build the Canal of Cuemanco for the Olympic games; in another report the *ejidatarios* of San Francisco Culhuacán are demanding compensation for land taken away in 1966 for housing purposes; as late as February 1992, 34 years after the expropriation of the Santa María Ticomán *ejido* for the Zacatenco campus, 26 *ejidatarios* are claiming their unpaid compensations.[32]

Uncertain payments and the possible total loss of control over the funds by local *ejidatarios* when the money is put under the control of FONAFE further reduces trust in the official process. The FONAFE fund has often over the years been accused of gross mismanagement by the *ejidatarios*. The *ejidatarios* have therefore every good reason to dread government expropriation, and strong incentives to sell the land on their own, even if that means selling the land illegally on the informal market. Informal markets for *ejido* land pay much lower prices than would be obtained in the primary market and have higher risks, but are still more attractive than expropriation.

Illegal conversion of *ejido* land for irregular settlement purposes becomes far more understandable within this framework, as a defense mechanism of the *ejidatarios* against expropriation, which is perceived by the *ejidos* as confiscatory. Once the meaning of *ejido* expropriation in the urban peripheries is interpreted in these terms, one can begin to understand appropriately *ejido* reaction to the regularization and reserve programs. However, some additional insights are still necessary to gain a clearer picture on the issues involved.

The Rationale for Discarding the ZUE Route in the Regularization Programs

The regularization programs were instituted not simply to remedy the tenure situation of the irregular settlements. As already mentioned, another objective, although one less advertised, sought to generate income for the agrarian establishment by using the regularization payments and the commercialization of any vacant remaining land in the regularized *ejidos*. Income extraction for the official agrarian establishment seems to be the basic reason behind the decision to abandon in 1973 the ZUE route as a valid method for legalizing the irregular settlements in the urban peripheries and shifting to the expropriatory route, after 16 years of upholding the validity of that legal mechanism.[33]

The reason for the shift is simple, the ZUE legislation exempts the ejidatarios from making any payments for their parcels. Furthermore, income obtained from selling any extra parcels to *avecindados* must be reinvested in the ZUE itself according to the agrarian code and can not be diverted to other *ejido*s as prescribed by the legislation then drafted for the expropriatory procedure. In other words, the ZUE route was discarded because it did not lend itself to generate income for the agrarian establishment while the new expropriatory procedure was well tailored for that purpose.

In the new procedure any settlement to be regularized was required to go through a formal expropriation process, in which the settlers would have to make payments to the regularization agencies for recognition of their property rights, regardless of any payments they might have made to the *ejidatarios* in the past to acquire their lots. These payments, assessed at a fraction of the land's commercial value, would have to be be made to CORETT, a nominally intergovernmental agency but which was basically linked to the agrarian sector and presided by the Secretary of the Agrarian Reform. Property deeds were to be given to the settlers once they completed their payments.

Assessment of the compensation amount due to the *ejidatarios* was to be determined on the basis of twice the agricultural value of the land, to be paid out to the *ejidatarios* according to the cash flow of the regularization program.[34] Although settler payments were substantially lower than the full commercial market price, they would still generate substantial income for CORETT and FONAFE. A very important federal property tax payment of ten percent on value was to be added to the expenses of the settlers, as required by law when real estate property changes hands.

The main income for the agencies would, however, come from the expropriation of still vacant *ejido* land during the regularization programs; this land could be sold at its full market price. CORETT thus became interested in increasing as much as possible the amount of vacant land expropriated from the regularized *ejidos* to sell it on its own. The procedure of *ejido* regularization, far from being a mere formality to cover legal requirements, was thus understandably turned into a tough confrontation between CORETT and the *ejidatarios* on the one hand, and CORETT and the *colonos* on the other.

A variety of issues, including the total area to be expropriated, the amount to be paid as compensation to the *ejidatarios,* and the price and terms of the resale operation to the settlers, were subject to intense bargaining. In the negotiating game CORETT was anything but an objective disinterested arbitrator between two conflicting sides, but, to the contrary, became an interested party trying to maximize income for itself by expropriating as much vacant land as possible from the *ejidos* for commercialization, paying as little as possible to the *ejidatarios,* and extracting as much as possible from the settlers. The *ejido* regularization process became established as a confrontation between *ejidatarios* , settlers, and government in which the *ejidatarios* were pitted against the settlers.

The *ejidatarios* were well aware that the only legal alternative open to them, that of expropriation, offered them nothing but the loss of their vacant land for an uncertain minimal compensation based on agricultural value. They therefore moved ahead of CORETT and preemptively sold their still vacant areas to irregular settlers as fast as they could to obtain the best possible deal in the informal market; these profits would be complemented in the uncertain future with any additional income that might accrue through regularization.

The clash of interests described for CORETT was even more pronounced for FIDEURBE and AURIS, which were established by the DF and SM governments as urban development agencies with far greater freedom of operation and clearly defined commercial objectives.[35] As a result, FIDEURBE became systematically involved in heavy conflicts with the *ejidos* in every regularization program until its liquidation in 1980,[36] as will be seen in the case studies presented in this chapter.

Failure of the Reserve

The power to carry out expropriation of vacant *ejido* land for new urban subdivisions was withdrawn from CORETT by the López Portillo administration in 1979, as previously stated, leaving this agency with the exclusive function of regularizing the existing irregular settlements.[37] The Ministry of Human Settlements and the DF government were designated in the 1984 amendments to the Agrarian Code as the beneficiaries of expropriations of vacant *ejido* land in the urban peripheries for establishing urban growth reserves, thus making it possible to actually fulfill the responsibility of establishing and managing land reserves for urban expansion that the Human Settlements Law had earlier placed on that Ministry.[38]

Whether stripping CORETT from its powers to expropriate vacant land was done in deference to the opposition of the *ejidatarios* as understood by Gilbert and Ward, or rather to transfer such powers to the then newly created

Ministry of Human Settlements may all be part of the answer. It is indeed ironic, as Ward observes, that the creation of the Ministry of Human Settlements in 1977 and the design to establish land reserves for planned urban growth, brought about the very opposite result and precipitated illegal sales of *ejido* land and intensified irregular settlement activity.[39] Nevertheless, within the general framework explaining *ejido* behavior developed here, this effect of the land reserve policy is fully understandable, and could have even been predicted.

As the old, easy approach of expropriating *ejido* land at agricultural value was left intact when the Ministry of Human Settlements became the beneficiary, the massive expropriations to supply the urban growth reserves were seen by the *ejidatarios* as far more menacing to their interests than the regularization programs. The reserve program threatened to leave the *ejidatarios* without land and without money while in the regularization programs they had at least benefited from their informal sales to the settlers. Stronger protests and intensified illegal sales activity were the response of the ejidatarios where expropriation was considered. Furthermore, as the takeover threat did not come from within the agrarian sector as before, but from an external political sector trying to penetrate the agrarian turf, implementation of the reserve now also faced the opposition of the agrarian political establishment as well.

It is now easy to see the reason why the reserve now suddenly awakened the opposition of the Ministry of the Agrarian Reform. As soon as the Ministry of Human Settlements and the DF government were designated as the beneficiaries of the expropriations of vacant *ejido* land for the urban reserve, CORETT and the Ministry of the Agrarian Reform stood to gain nothing from reserve expropriations. On the contrary, every hectare of land expropriated in anticipation of growth diminished the need for regularizations. On the other hand the regularization programs in which they were the designated beneficiaries had become an important steady source of income.

The conclusion is clear, *ejido* resistance to government takeover is not the result of a situation of political privilege of the *ejidatarios*, but to the contrary originates in a set of inequitable institutional arrangements which ignore their property rights. With rare exceptions, one will find that little attention has been paid in the literature to the claims of the *ejidatarios* and the legal and moral basis on which these claims are made.[40] Most authors confuse the *ejidatarios* with the official agrarian establishment and have gone along uncritically with the implicit assumption that their claims are easily disposable. This attitude is rooted in a perception of the *ejido* land as state owned, which the state has the right to reclaim at the lowest price possible, and of the *ejidatarios* as tenants lacking property rights. This extended view will be challenged in the next chapter as untenable on legal and historical grounds. Three case studies will then be presented in chapter V which reflect the different issues and situations discussed in this and the next chapter supplying additional evidence for the explanations developed in this chapter.

---

1 As *ejido* and *comunal* land are subject to almost the same restrictions, the term *ejido* is normally used to refer to both of them, unless there is a need to differentiate.

2 Ley Federal de la Reforma Agraria. L.4. Art 267.

3 (Ibid. art. 112-125).

4 (Ibid. Chap VIII, art. 112)

5 (D.O. January 17, 1984).

6 (Ley Federal de Reforma Agraria, art. 112-VI; LFRA).

7 (Ibid. art. 122-I, II, III).

8 (Ibid. art. 122, I-II).

9 (Ibid. art. 122-125).

10 (Ibid. art. 112-VI; art. 122-II).

11 (DO January 17, 1984).

12 (Ma. S. Cruz Rodriguez, 1981).

13 (Schteingart, 1978:p. 74, 1983; García, 1982).

14 (Garcia Peralta, 1983).

15 (Ward, 1981: p.45).

16 (Schteingart, 1983: p. 293. Iracheta, 1984).

17 (Stephens R., 1981; PDUDF p.165-66).

18 (Ward, 1981:p. 48-53; Gilbert & Ward, 1985:p. 89-91. Schteingart, 1981; Iracheta, 1983; Ann Varley, 1985:p.5, 10-11.) Many more authors such as Tomasz Sudra, Patricia Connolly, Martha Schteingart and Alfonso Iracheta have dealt with this association.

19 (1985 :p. 89-91; Ward, 1981:p. 51-53).

20 (1986: p. 90)

21 (Ward, 1981: p. 53; Gilbert & Ward, 1985: p. 90-91).

22 Ibid.

23 Ibid; (Ward, 1981: p. 52;).

24 Ibid.

25 Ibid.

26 (COPEVI "Las Politicas Habitacionales del Estado Mexicano," p. 138,1977).

27 According to An Varley (1985), one third of all the ejidos subject to regularization between 1970 and 1982 asked for court injunctions against the expropriation decrees.

28 "There is a very clear reality here in the capital: the value of one square meter here is higher than an hectare of land in the rest of the country" commented a high local government official when discussing the irregular settlements on ejido land in the MAMC periphery (NYT, Rother q. Merino, July 15, 1987).

[29] Confiscation is defined as the taking of property without compensation, and is strictly forbidden by the 1917 Constitution.

[30] (DO September 25, 1945).

[31] (Eckstein, 1977: p. 64).

[32] Excelsior, February 22, 1992.

[33] (LFRA, art. 112-127).

[34] The total process of regularization of a settlement was estimated by CORETT to take an average of 9 years (Síntesis sobre el Programa, CORETT, 1980; Stephens R. 1981).

[35] FIDEURBE was empowered within the DF to further the expropriation by the federal government of a) irregular settlements, b) *ejidal* and *comunal* lands of low productivity, c) decadent areas subject to rehabilitation or remodeling. It was then allowed to sell that land to the DF government, the federal and local housing agencies, and to the residents of the irregular settlements.

[36] "The head of Fideurbe complained that it was almost impossible to carry out regularization effectively owing to the constant stream of protesters and petitions passing through his office," (Gilbert & Ward, 1985:p. 191).

[37] (Diario Oficial, April 3, 1979).

[38] (DO January 17, 1984; LFRA, art. 117).

[39] (Ward, 1981:p.5.).

[40] Francisco Covarrubias Gaytán, former undersecretary of SEDUE at the 1981 SIAP/SMP Tlaxcala Conference (chapter V): "The *ejidatarios* are affected in the expropriation process by the small compensation payments, the protracted administrative procedures for their actual payment, and above all by the loss of the resource that furnished them with a living; it is fundamental that the incorporation of *ejido* land to urban development should contemplate the interests of the *ejidatarios* with social justice (SMP, 1982: p.16)."

Chapter IV

THE *EJIDO* PROPERTY RIGHTS: AN HISTORIC OVERVIEW

The literature has shown an understanding of the problems and demands of the irregular settlers, but one does not find such an understanding of the *ejidatario* viewpoint. The *ejidatarios* implicitly come out in the literature as the culprits of the pervasive illegality that characterizes the urbanization process in the MAMC periphery, a notion which may look true, but which in a deeper sense represents a distortion of reality, as was seen in the previous chapter.

As overwhelmingly important as the needs of the great masses of urban residents and settlers are, such a lack of understanding for the *ejidatario* viewpoint does not allow to see the problem of land conversion in the periphery of the MAMC in the proper light. If there is any lesson to be learned from the aborted attempts of the past to exercise some measure of control over urban growth in the MAMÇ, it certainly points to the need for taking fully into account the property rights and interests of the *ejidatarios* no less than those of any other interests.

A struggle for land is taking place in the urban periphery which has much in common with the land usurpations of the historical past. The *ejidatarios* are losing their land and agricultural livelihood, although now it is not to the *hacienda* but to the city. Even if only for that reason, it is necessary to seek a fuller understanding of the *ejido* institution and to take a closer look at the historical roots and legal and moral rights of the *ejido*, in order to gain a fuller and more balanced view of the issues involved.

Such a discussion is particularly relevant when dealing with the meaning and scope of the property rights of the *ejido* communities in general, and the compensation issue in particular. The great forward thrust of statism in Mexican political and economic life gave rise to efforts to reinterpret *ejido* property rights at the threshold of the seventies as usufruct rights given by the government to tenants on national land, as if the *ejido* were a type of state farm as is the case in the Soviet system. It is felt that such an approach, which looks at *ejido* property as if it were government property, not recognizing its true nature, is in the background of the permanent struggle around *ejido* land, including the frustrated attempts to establish an urban reserve.

This statist view, which has been given ample circulation inside and outside of the political arena, requires close scrutiny due to its important consequences. It intentionally confuses **original property by the nation,** which applies to all types of land, with state property at a specific moment in time, which does not apply to the *ejido* or to private property. The statist view serves the purpose of ideologically weakening the *ejidatarios'* claim to their land vis-a-vis the state.

The statist view has found its way to the national and international social-science literature without any real basis in the history and jurisprudence of the institution, and has been uncritically taken for granted by such authors as Castells,[1] Ward, Schteingart, García Peralta and others. One thus typically finds the following definitions: "*Ejido* land is property of the state given in concession of usufruct;"[2] "*Ejidal* land is legally the property of the state. Land use rights are transferred by special concession to entire communities;"[3] "The *ejido* is constituted by land which belongs to the nation that is handed over in usufruct to a group of peasants;"[4] "Land held by the community over which peasants have usufruct rights only."[5] It will be shown in this chapter that such a view is untenable and is disproved by all that is known about the ideology of the agrarian reform movement during and after the revolution and the way these principles were expressed in the 1917 Constitution and later interpreted in the earlier agrarian legislation.

In a wider perspective this view can be related to the general unwritten national policy of maintaining the *ejidatarios* in a state of permanent tenure uncertainty, withholding from the majority of the individual *ejidatarios* the titles that give them permanent possession of their assigned agricultural plots as required by the 1942 Agrarian Code.[6] It likewise fits in well with the ample powers given since 1971 by the Law of the Agrarian Reform to FONAFE to divert the proceeds of urbanization from the *ejidos* being expropriated to the economic development of other *ejidos* throughout the country.

A clarification of the meaning of *ejido* property rights is crucial to any serious effort to implement urban land reserves in the future, with which the last chapter of this book is concerned, as mandated by the Constitution since the amendment of Article 27 in 1976. It has been cogently argued in the previous chapter that the failure of establishing the urban reserve until today is very much the result of not facing the question of *ejido* property rights frontally and of assuming that political manipulation of the *ejidatarios* and token compensation will somehow achieve such an outcome.

The attempts to reinterpret *ejido* property rights as state property are implicit in the wider national discussion opened in more recent times at the closing of the IV Reunion of the Republic by President López Portillo in 1980. In answer to the critics of the Law of Agricultural Development (Ley de Fomento Agropecuario), President López Portillo opened a national debate on the convenience and feasibility of turning all land property rights in general into public property, bringing to an end private and *ejido* property within the framework of the Mexican political system. The short-lived debate elicited a series of responses from the most prominent Mexican jurists, such as Lucio Mendieta y Nuñez, Andrés Serra Rojas, Antonio Martínez Báez, Alberto Trueba Urbina and others who categorically expressed their view that such an action could not be accommodated within the 1917 Constitution. It would, they said, constitute a break with the legal and political institutions of the country, amounting to no less than a revolution.[7]

This question of the *ejido* property rights in the urban peripheries can not be understood outside of the framework of the history and ideological discussions about that institution. *Ejido* property rights have undergone changes in form and content over time, which bear heavily on every aspect of their contemporary interpretation. One must, therefore, survey the transformations of these rights within the general context of the history of the institution to shed some light on the enormous complexity of the land issue in the periphery of the expanding cities. This is even more important where the objective to be attained is the formulation of policy changes as is attempted in the last chapter.

THE PRE-COLUMBIAN AND THE COLONIAL *EJIDO*

The *ejido* institution in general has an ancient history which goes back to pre-Columbian times, when the Indian villages were the collective owners of their land. Many Indian villages in Central Mexico based their property claims, during colonial and modern times, on pre-Columbian possession. The Spanish crown gave recognition for the first time to the right of the Indian villages to own land in a decree by Carlos V in 1533. The land being distributed to the *conquistadores* was not be taken from those held by the Indian villages. This decree complemented the earlier land distribution law ("Ley para la Distribución y Arreglo de la Propiedad") signed by King Fernando V in Valladolid in 1513, which served as the basis for the distribution of land in the newly discovered continent, which Spain had set out to conquer.[8]

The land distribution actions which the colonial government undertook under the legislation mentioned produced four forms of rural landholdings, which determined the character of Mexican rural life for centuries: the *haciendas*, the large landholding estates which were granted as private property to the *conquistadores* and their descendents by the Spanish crown; the *ranchos*, the smaller private rural properties, well under 100 hectares in size, which were worked by the owners and their families, and a few sharecroppers or hired laborers; the communal lands of the free Indian villages which operated under a combined collective and individual work system and Church property.

Church property received full recognition in colonial law. The property rights granted to the Church were not of a individual nature but rather of a corporate nature similar in this particular point to those of the free Indian villages. The Church became a very important holder of landed property during the colonial period in the rural areas of Mexico, as a result of direct endowments by the Crown for its missionary activities, the gifts and inheritances that it received from private property holders and its mortgage lending activities. Thus many of the large estates during the colonial period up to the days of Juarez were owned by the Church.

Specific legal form was given to the land rights of the Indian villages by the "Consejo Real de las Indias", the administrative authority over the Spanish colonies in the Laws of the Indies. In the decrees signed by King Felipe II on May 26, 1567, and on December 1, 1573 the lands of the Indian villages were declared to be inalienable, inextinguishable, and unmortgageable. The land belonged in a corporate manner to each village community, not to its individual members. The lands of the free villages comprised four types of property: The *Fundo Legal*, the *Tierras de Común Repartimiento*, the *Ejido* and the *Tierras de Propios y Arbitrios.*

The *Fundo Legal*, the forerunner of the present day *zona urbano ejidal* (ZUE), included the residential area where the village proper was located with its blocks of houses and individual parcels. The physical dimensions of the *Fundo Legal* was defined a little more than a century later, by another royal decree (July 12, 1695), as covering an area of 600 varas, the equivalent of a square having 4190 m. on each side, positioned in the center of the village.

The *Tierras de Comun Repartimiento*—a term which has survived until the present when speaking of *comunal* property—refers to the land that the Indian villages were allotted in a corporate form by the colonial government to be distributed individually among the villagers for their sustenance. These parcels are the historical forerunner of the individual agricultural parcels of the modern day*ejido* created by the Mexican Revolution. The rights of the individual villagers to exploit these parcels was permanent, unless they ceased to cultivate them for three consecutive years, or were absent for the same period, in which case the community would assign them to other members. The communities guaranteed that this land would actually be distributed and cultivated by the peasant families, which in their turn were expected to engage, in addition to subsistence production for their own needs, in providing seasonal labor to the *haciendas.*

The *ejido*, the third component of the village property, was a form of communal land property for the collective non-agricultural free use of the village. It is rooted in the Roman *ager publicus*[9] and its name is derived from the Latin *exitus*, exit in modern English, referred to as a relatively limited portion of land bordering on the village—in its "exit" to the fields so to speak—which was not cultivated, but was used in common by the villagers in Spain as a threshing floor or as a meeting ground of their herds. This was the original meaning of the *ejido* in Spain, which in colonial Mexico was modified and expanded to include the woods, pastures, and waters used and exploited collectively by the villages.[10] In modern times due to the imprecise use of the term by Luís Cabrera in 1912 which used the term to refer to the individual agricultural parcels belonging to the *comun repartimiento* category, as then critically pointed out by Wistano Luís Orozco, it lost its precise original connotation and came to designate the village community as a whole and/or its agricultural plots.[11]

The *Tierras de Propios y Arbitrios* was the land used collectively by the Indian villages to obtain income for the community to maintain their local government, cover the expenses of their religious festivities and pay taxes to the colonial government. It was the common practice of the free villages to lease these lands to neighboring *ranchos*, to obtain cash. This land category which played a crucial role in maintaining the village autonomy was not reestablished by the Mexican Revolution.

A symbiotic relationship was established between the *haciendas* and the *ejidos*. The *haciendas*, although self-sufficient in almost everything, badly needed the labor that the villages provided. The *haciendas* grew all the food they needed, and produced their own tools, building materials and the other components of agricultural operations. The free Indian villages, for their part, held control of land for their subsistence and were even more self-sufficient than the *haciendas.* These landholding forms in the farming areas of Mexico kept the rural population strongly tied to the land and isolated from the commercial system of the cities and coastal towns.[12]

Although the history of the colonial period is plagued by conflicts over property limits between the Indian villages and the neighboring *haciendas*, the Indian villages managed to retain collective property rights on much of their land throughout the colonial period under the Spanish Crown. After independence, however, the land conflicts between the Indian villages and the *haciendas* took a much more violent form and brought about such tragic events as the War of Castes in Yucatán during the Mexican-American war, in which the Mayan Indians were deported and sold as slaves in Cuba, while the hemp-producing *haciendas* grew on the *ejido* lands of the villages.[13]

THE LERDO LAW: ABOLITION OF CORPORATE PROPERTY

The liberals in independent Mexico adopted in their political platform, shortly after the Mexican War of Independence in 1821, the goal of ending corporate landholdings held by the church and the village communities, as a means to opening them up to commercial development. Spanish liberals had already advocated in the eighteenth century, before the War of Independence, the conversion of the lands held by the peasant communities to private property that could be bought and sold, as well as mortgaged. The liberals envisioned great economic gains and argued that peasants who would become owners of their lands would have new incentives to increase production. The issue was turned into a widely debated national question and, as a result, many individual state governments in the young republic adopted legislation to end community landholdings from the late eighteen forties on.[14]

The major onslaught on communal village property took place when the liberal faction finally came into power in the middle of the 19th century. After the last conservative government of Antonio López de Santa Ana was ousted by the liberals in 1855, the triumphant liberals finally made privatization of corporate property into national law on June 25, 1856. President Juan Alvarez, backed by Finance Minister Miguél Lerdo de Tejada, and Justice Minister Benító Juarez adopted the *Ley Lerdo*, named after its author Miguél Lerdo de Tejada. This law, also known as the *Ley de Desamortizacion*, abolished nationally the property rights of all corporate organizations such as the Church or the village communities, and recognized only physical individuals as being legally capable of holding title to property. Some months later on February 5, 1857, the Lerdo Law became part of the Mexican Constitution.[15]

The *Ley Lerdo* did not directly expropriate the lands of the Church or peasant communities, but prescribed that their holdings were to be sold to the current tenants at prices based on existing rents, and that these payments were to be extended over a 20 year period. Only properties not claimed by their occupants would be auctioned off. The income obtained from these mandatory sales would go to the former owners, Church institutions or village governments, while the state would gain a tax on each transaction. However, as things turned out, when the large estates of the Church and property held corporately by the Indian villages were actually divided, the tenants very often did not become the owners, but title passed to other parties.

The privatization of the community lands produced more losses than gains for the villagers. True, the implementation of the Lerdo Law turned the villagers into the private owners of the *repartimiento comun* plots which they worked individually. However, as far as the *Tierra de Propios y Arbitrios* is concerned which they leased out to neighboring *rancheros*, the new law led to their being auctioned off to the nearby *haciendas* and *ranchos*, enlarging those properties. The *ejido* lands, the collective woodlands and pastures, which were often located at considerable distance from the villages, became the property of the municipalities.

In the overall balance, the villages came out the great losers, and they were well aware of this outcome from the very beginning. As soon as the liberals announced their reforms in 1856, the majority of peasant villagers opposed these reforms. They understood that they would not gain any additional lands to those they already held as *repartimiento comun* plots, and would lose ownership over the other two types. Their access to the woodlands and pastures would be cut off, and they would forfeit the income from the leased lands which made possible the autonomy of their village government. The fears of the villagers came true, and while many of them emerged as individual owners of small plots of land, they lost the access to the pastures and woodlands they previously enjoyed, and their village government's independence and power. The villagers across central Mexico protested vehemently, and armed resistance blocked reform in several areas.[16]

Very soon after the proclamation of the Lerdo law in 1856, intensive privatization of village lands took place in communities near the larger cities. Such early implementation was uncommon away from urban centers. Near the cities, many village community leaders agreed to rapid sales of properties, under the pressure of powerful tenants with political backing. Once these sales were negotiated, many of the villages faced continuing difficulties in collecting the payments due to them. The liberal leaders appeared more concerned with privatizing community properties than with insuring that the villages obtained the payments. These losses no doubt increased many villagers' opposition to the privatization drive.

Among the *hacendados*, there was little protest against the alienation of community lands, which they perceived as serving their interests. They understood that the reform would not only make it possible to increase the size of their landholdings, but would also bring about greater availability of scarce labor from the declining landholdings of the peasant communities, making *hacienda* operations more profitable.

The privatization of Church properties which deprived the church of its urban and rural properties, by contrast, brought about a deep split in Mexican society. The conservative elites joined to defend the interests of the Church. The differences between the two factions, the liberals and conservatives, brought about a civil war, which lasted for 10 years, in which the conservatives unsuccessfully attempted to establish a monarchy, until their defeat in 1867. As a response to their opposition, President Benito Juarez decreed the nationalization of Church properties not yet privatized in 1859 to be auctioned off to the highest bidders, with the proceeds going to the liberal government's treasury.[17]

During the decade of the civil war, the privatization of community lands ceased completely. However, as soon as the monarchy fell, and the conservatives were defeated, the liberals used their unopposed power to resume implementation of the Lerdo law in the central highlands. Elsewhere, implementation began slowly. As a result, villagers in the central regions of Mexico rebelled in the late sixties against the privatization of their lands, and even brought forward demands for the redistribution of the *haciendas*.

Following the defeat of the conservatives, their potential political allies, and the alienation by the liberals, the villagers, who had lived isolated for centuries, became receptive to urban radical leaders. An ideology of agrarian rights began to develop among a small group of radical intellectuals in the late eighteen sixties during the uprisings in the central highlands. They proclaimed the right of the peasants to subsistence lands linked to autonomous communities. Many villagers showed willingness to work with these leaders and were ready to fight for that ideology.[18]

Widespread rural insurrections broke out.[19] Julio Chávez López, a Fourier socialist and forerunner of Zapata in the State of Morelos, organized a rebellion during the presidency of Bénito Juarez and was tried and executed in Chalco on September 12, 1869.[20] In Chiapas, the Tzotzil Mayas rebelled in

1868. In the North, the Yaqui Indians of Sonora, who had been deprived of their lands and forced to divide their communal property by the landowners, initiated a bloody rebellion in 1873, during the term of liberal Vice President Sebastián Lérdo de Tejada (1872-76), a brother of the author of the Lerdo Law, who became president after Juarez' death in 1872. This was the first of a long series of Yaqui uprisings which lasted for many decades.[21]

Although the Lerdo Law and the Constitution of 1857 stressed economic liberalism and free enterprise, and sought to create an enterprising rural middle-class, it actually led in the days of the Porfirio Díaz government to a very different result in this respect. The early well-meaning liberals of the Juarez period hoped by means of the Lerdo laws to create a class of independent prosperous small freeholding farmers in the U.S., English, and French tradition. The unintended result, however, was not a rural middle class, but monopolization of the land by the *haciendas*, and polarization at the extremes: a small class of wealthy *hacendados* versus large masses of virtual serfs.[22] One can cite this outcome as one more striking historical instance of the justified fear that the Scottish moral philosophers of the eighteenth century felt at the time of the French Revolution for the unintended effects of well-meaning policies.

THE DÍAZ ERA: THE USURPATION OF THE PROPERTY RIGHTS OF THE *EJIDOS*

With the ousting of Sebastián Lerdo de Tejada by Porfirio Díaz in 1876 through a coup d'etat, the long Porfirian era (1876-1911) of capitalist development and dictatorship began. It brought the agrarian issue to its crisis. Although Porfirio Díaz claimed to be a defender of village interests, his policies are linked to the increasing alienation of village lands.

Díaz adopted a new law in 1883, known as the *Decreto de Colonizacíon y Compañías Deslindadoras*, to promote rapid colonization of the sparsely settled northern frontier. This law which laid down the norms for the survey, subdivision, settlement, sale and privatization of vacant and national lands. Although the *Ley de Colonizacíon* contained limitations on the type of land that could be appropriated in the process and the maximum size allowed to individual landholdings, it became a new source of usurpations, increasing property concentration and monopolization. The survey companies could lay claim on land and the burden of proof concerning ownership lay with occupants. The *haciendas* seized land of the Indian villages which had survived the Lerdo law as well as land of private owners. The largely unenforced limitations on this law were eliminated in 1893 to reward political favorites and strengthen the hold of its political machine on the country in the *Ley sobre Enajenación de Terrenos Baldíos*.[23]

The effects of the Lerdo Law and the Colonization Law during the Díaz era can be better understood against the background of the economic policies pursued by that government, which led to the concentration of land property in large estates. The Díaz government was occupied in laying down the basis of a modern capitalist industrial economy. The framework of an industrial society was being rapidly developed: transportation, communications, mining, and to a smaller extent, manufacture. This effort was mostly based on foreign investment and external markets.

Díaz built one of the largest railway grids on earth to integrate the Mexican economy and link it more closely to the United States.[24] The railroads brought about a reorientation of much of the nation's mining and agriculture towards export, mainly to the United States. By linking urban and mining centers with the Mexican coastal cities and the US frontier, the railroads facilitated a dramatic growth of exports in the mining sector to supply the large iron and steel plants of the United States, and also increasingly in the agricultural sector, to the growing U.S. market.[25]

New opportunities for commercial export of agricultural products and cattle developed rapidly in the areas served by the rail network. The large estates increasingly turned away from their traditional feudal system of subsistence agriculture to the production of staple commodities for export, employing seasonal wage labor and sharecropping.[26] As a result, every new rail project triggered a wave of communal and small property land usurpations in the areas served by the lines. The lands held by villagers and small holders in these areas became the target of usurpation, while the courts repeatedly backed the claims of the powerful, whatever their legality.[27]

Over two million acres of *comunal* land belonging to the villages were privatized during the Porfirian period, which ended up in the hands of estate owners or of land companies.[28] Privatization on the basis of the Lerdo Law or the colonization laws became often an excuse for confiscation. With disregard for the rights of the occupants, on whom the new laws laid the burden of proof concerning ownership, whole rural communities in the central region of the country which could not prove their ancient rights were dispossessed of lands that they and their ancestors had worked for centuries.[29]

A point was reached where Díaz himself took legal steps to slow down the loss of the village lands. He amended in 1901, Article 27 the Constitution of 1857, allowing non-church corporate bodies such as the Indian villages to hold lands.[30] Nevertheless, it was too late to turn the clock back to undo the harm done. By 1910 some 90% of the villages in the central plateau, and 85% in the rest of the country, had lost their land. 90 % of rural families in the entire country were landless and fully 50% of the rural population became tied to the *hacienda* system.

The reorientation of the agricultural sector towards commercial export production was one of the major factors that brought about the ruin of the Porfirian political system.[31] The estates ceased to produce enough food for internal consumption to feed the population.[32] As food became scarce and prices rose, Mexico, an agricultural country, had to import grain to feed its population, pushing the peasants toward starvation. The reaction of the peasantry to the loss of their lands and starvation was quick in coming, principally in the states surrounding Mexico City such as Morelos, Puebla, Tlaxcala, Mexico, and the southern part of the Federal District.

THE MEXICAN REVOLUTION: REESTABLISHMENT OF THE *EJIDO* PROPERTY RIGHTS

Political elites dissatisfied with the Díaz regime began to plot the overthrow of his government. Demands for greater political democracy, expressed in the slogan, *sufragio efectivo no reelección,* "effective voting, no reelection" kept together the political and economic elites which brought to power Francisco I. Madero. For many who joined the struggle against Díaz, however, demand for agrarian reform, not democracy, was the main issue. Such was the case with Zapata, who assumed leadership of the agrarian discontent in Morelos in 1909 under the banner of land and liberty, and proclaimed that "the land belongs to its tillers." Andrés Molina Enriquez, an intellectual who published in that same year his book *Los Grandes Problemas Nacionales,* one of the basic ideological documents of the Mexican Revolution, called for the division of the large estates, and the restoration of the system of communal ownership.[33] Madero, far more focused on political reform than in radical land reform, nevertheless included in article 3 of his revolutionary proclamation, known as the Plan de San Luís (October, 1910), promises of restitution of the village lands. For him and many of his followers such a reform meant the return of the lands that had been unlawfully taken away from the peasants:

> Abusing the law of vacant lands, many small owners, mostly native Indians, were dispossessed of their plots...As justice demands the restitution of the lands that were taken away in such an arbitrary fashion, the decisions and verdicts are declared subject to revision,

> and they who acquired them in such an immoral manner, or their heirs, will be required to return them to their original owners, to whom they will also pay compensation for the damages inflicted.

When Díaz was finally ousted in 1911 and it looked as if the political objectives of the revolution had been achieved, it did not take too long to realize that the agrarian problem would not allow the country to regain its peace. The coalition of interests that had succeeded in bringing down the dictatorship was starting to show its inner divisions, particularly on the agrarian issue. Dissatisfied with Madero's approach and initial slow pace on the agrarian question, Zapata broke with Madero a few weeks after his installation and accused him of not fulfilling the promises of the *Plan de San Luis*. On November 25, 1911 Zapata issued the *Plan de Ayala*, in which he not only demanded the return of all the appropriated lands to the villages, but also advocated the distribution of one-third of the *haciendas* among the peasants, and the expropriation and distribution of all those *haciendas* whose owners had fought against the revolution. Former allies turned bitter foes, and soon the federal army was confronting the Zapatista guerrillas.

The Madero ministerial cabinet and the Congress were considering in 1912 a restitution project to return to the villages the lands that were unlawfully taken away from them. They were also considering small land endowments as a means to supplement the peasants' salaries during the months that they were unemployed at the *haciendas*. Luís Cabrera,[34] whose ideology dominated the agrarian reform until the days of President Cárdenas, made the following statement to the Chamber of Deputies on December 3, 1912, in defense of his Agrarian Law Proposal :

> The rural population needs a complement to its salary: If they would be in possession of *ejidos*, they would work half a year as laborers, and in the other half they would devote their energies to harvest on their own. Not possessing them they are forced to live six months on their salary and the other six months they take the rifle and become Zapatistas.
>
> As long as it is not possible to create a system of agricultural production based on small property to take over the place of the large producing estates, the agrarian problem must be solved through the production of the *ejido* as a complement to the laborers salary.

With Madero's assassination (February 22, 1913) and the escalation of the civil war, the agrarian issue gained greater significance. Two doctrines in addition to the Cabrera proposal on the mini-*ejidos* began to crystallize among the revolutionary fractions. One approach advocated by the northern revolutionaries, loosely described as the "Northern view," believed in the "right of each man to own, in fee simple, a plot of land." They held an evolutionary liberal approach to the agrarian question, and shared with the liberals of the previous generation the vision of creating a rural middle-class of independent small farmers. Cabrera himself also shared this view and visualized the restoration of the *ejido* as a first step to small private rural property.

According to this view, once the villages were endowed by the government with land from the nearby *haciendas,* through such means as purchase, expropriation with full compensation, mandatory renting or sharecropping, the land would finally be subdivided into private family landholdings, to coexist with the large landholdings on equal terms,[35] duly protected by law from being swallowed up by them.[36]

The other more radical view held by Zapata and his followers in the South is known as the "Southern view." For them the restoration of the *ejidos* meant a return to its original historic roots and the reestablishment of communal land tenure on a permanent basis. Zapata stressed the Indian past of the country and the intimate relationship of the *ejido* to this past. His vision of the *ejido* meant the reestablishment of village collective property and self-government, whereby property of land, distribution of parcels, decisions on production, collective work, and mutual help were decided on by the community.[37] Antonio Soto y Gama, one of Zapata's closest advisors, described these two differing views:[38]

> The Northerner was more individualistic, more removed from the old concept of the communal calpulli, more desirous of fully exercising the role of an independent proprietor, demanding for himself a portion of regular sized land which would be totally his and under his complete control, without the restrictions or taxes that the traditional indigenous communal structure imposed; therefore instead of asking for the revival of this traditional structure, as the Southerners desired, he aspired to be able to exploit and extensively cultivate the lot of land assigned to him in the land distribution, including the right to sell, to transfer, to mortgage, to impose taxes for the acquisition of funds or to contract loans as required.

Zapata actually implemented his version of land reform, in the short period that the military situation allowed, in the Commune of Morelos. His followers reestablished their local government, system of justice, and traditional productive organization. After entering each village they systematically destroyed the archives of the Public Registry of Property, and actually distributed all the 53 *haciendas* of the state, the ranchos, and farms. They resumed production in the haciendas, and put into operation the sugar mills and refineries under a collective system.[39] Unlike the later agraristas under President Cardenas, the zapatistas envisioned their peasant communities as independent from the state and allowing full freedom to its individual members.[40]

Thus, three different visions of agrarian reform, which were never reconciled, coexisted from the outset of the Mexican revolution: the Northern view, which advocated adequately sized small private farms, *ranchos*, for independent farmers capable of providing full sustenance to the farmer families; the Southern view, which envisioned a return to communal village property and production; and the Cabrera view, which sought an interim solution based on the *minifundio*, the mini plot belonging to the communal *ejido*, to be used as a complement to salaried income in the hacienda. As pointed out by Clarence Senior, one result was an attempt to act upon all three ideologies with divergent results.[41]

Carranza, the flag-bearer of constitutional legality who fought against Villa and Zapata, emerging as the leader of the wining faction during the revolution, included in his revolutionary program, the *Plan de Guadalupe* (December 12, 1914), an agrarian reform program to win peasant support. He promised not to wait until the revolution is over, but to issue laws while fighting was going on, to return to the villages the lands that were unjustly taken away from them, to break up the haciendas, and to favor the establishment of the small rural property. The first step in this direction was his publication of a decree on January 6, 1915,[42] whose main author was Luís Cabrera, his advisor, which declared null and void all the transactions that illegally deprived the villages and communities of their "lands,woods and waters" since the adoption of the Lerdo law of June 25, 1856.

True to the liberal view, the return of the communal lands taken from the Indian villages was seen in this decree, as a step leading to the establishment of small family holdings. Its commitment to the dual purpose of the agrarian reform and of maintaining that reform within the boundaries of private property as understood by the Northern liberals is clearly expressed in the text of the 1915 decree: [43]

> Not to revive the ancient communes nor to create others similar to them, but only to give the land to the miserable rural population that lacks it today, in order that it may fully develop its right to life and liberate itself from the economic servitude to which it is now reduced; moreover, it must be noted that proprietorship of the land will not be vested in the commune of the pueblo, but will be parcelled out in full dominion with only such limitations as are necessary to

prevent avid speculators, especially foreigners, from monopolizing such property.

Aware that restitution of lost village lands was only part of the problem, and that demand for land by landless peasants was the more important motive behind peasant unrest, the decree also opened new ground and established that endowment on the basis of need could still take place where villages could not show title to the lands to which they claimed to have been deprived. The 1915 decree, which later was incorporated into the 1917 Constitution, also established that lands indispensable to communal groups should be provided at government expense. The decree brought about in the beginning years *ejido* land grants of greatly reduced size, which very often came out to less than one hectare per peasant, within the guidelines of the Cabrera concept.[44]

A National Agrarian Commission to administer the new law was mandated by the Carranza decree.[45] Well aware of the great power that the implementation of this decree was putting into the hands of the generals, chieftains, and local politicians, Carranza made public an administrative presidential resolution on September 19, 1916, denying state governors the right to distribute these lands. As the Carranza movement ultimately triumphed over the other factions, this decree became the legal basis of agrarian reform.

Carranza believed that there was a need for a new Constitution to bring about the structural changes that the country required. The day he stood up in arms against Huerta as a result of the assassination of Madero and the overthrow of his legally established government, he had expressed this need before an audience in the city hall of Hermosillo, Sonora, on September 24, 1913. In late 1916 he convened a Constitutional convention to reform the Constitution of 1857 in Queretaro.[46] Agrarian reform was one of the main issues to be legislated at the convention.

THE QUERETARO CONSTITUTION AND THE AGRARIAN REFORM

The Constitutional assembly, in which the dissident revolutionary factions of Zapata and Villa were not included, was convened in the historic town of Queretaro. The convention lasted for two months, from December 1, 1916, to January 31, 1917, and produced within that short period a new Constitution for the country, known as the Queretaro Constitution or as the Constitution of 1917. The new constitution was promulgated symbolically on February 5, 1917, 60 years to the day after the 1857 liberal Constitution had been officially proclaimed.

The reform of Article 27 of the Constitution of 1856, which was intended to lay the legal basis for land reform, is considered the most distinctive feature of the Constitution of 1917. This article, which was formulated in the last ten days of the Convention by a special committee to which Andrés Molina Enríquez acted as advisor, raised the decree of January 6, 1915 to Constitutional rank, and departed from the 1857 Constitution and the subsequent Díaz legislation in various important points. Its most important features are the restitution of the communal lands lost by the villages; the establishment of the right to new land endowments, *ejidos*, to peasant communities applying for land; the mandatory subdivision of the large estates above a maximum size;[47] and the cancellation of the private property rights to the subsoil and to its free exploitation, which the Díaz legislation had granted to the private owners in the laws of November 22, 1884, and June 4, 1892 in order to promote mining activity.

The framers of the new Constitution adopted the thesis proposed by Andrés Molina Enriquez, that it was necessary to return to the basic feature of colonial property law: the king's right to "absolute ownership."[48] This right had passed to the Mexican nation as the successor of the Spanish Crown. The nation as the king, had the right to impose on private ownership such limitations as the public interest required. The nation could retain under its domain as much as the national interest required, and could grant property rights to whom it considered appropriate. The introductory paragraph I of Article 27 reaffirmed the principle of eminent domain and defined private property as a right that the nation grants to private persons:

> Ownership of lands and waters within the boundaries of the national territory is vested **originally** in the Nation which has had, and has, the right to transmit title thereof to private persons, thereby constituting **private property**.

Wishing to recover for the state the control of the mineral resources that the Díaz Government had opened up to private foreign enterprise, the committee made a distinction between surface and subsoil property rights. Like the king, who could dispossess the grantee of private property and return it to the royal patrimony if necessary, the nation had the power to recover full rights over the minerals and waters that were being held as private property. The nation would thus recognize or grant to private persons direct ownership of the surface of the land under the "same conditions in which it was exercised by private persons during the colonial period, and later recognized by the republic", but would "retain under its domain the products of the subsurface, not bestowing upon private persons more than the exploitation rights of these properties," considering that "cession had not included the right to the products of the subsoil." These concepts were clearly expressed in the Constitutional text of article 27.

> 4) In the nation is vested direct ownership of all minerals or substances which in veins, layers, masses or beds constitute deposits whose nature is different from the components of the land...;
> 5) In the nation is likewise vested the ownership of the waters of the territorial seas...; the waters of the principal rivers or tributaries..; etc;
> 6) In cases to which the two foregoing paragraphs refer, the ownership of the nation is inalienable and imprescriptible, and only shall concessions be made by the Federal Government to private parties or to civil or commercial corporations organized under Mexican law.

It is obvious that the Constitutional assembly had no intention of abolishing private property rights as far as the surface of land is concerned, but was rather interested in reaffirming the principle of eminent domain, as was clearly expressed during the discussions and attested on many occasions by Molina Enríquez: "the eagerness to abolish individual property, cannot be considered in its essence as anything but utopian."[49] It thus provided for the expropriation of surface rights when necessary, but emphatically rejected confiscation of property (Art. 22) and required that "expropriations shall only be made for reasons of public utility and by means of indemnification," as expressed in paragraph two of article 27.

The next question which the framers of article 27 had to face was that of communal property of land that had been abolished in 1856. They deplored that "the property laws in force ignore the fact that there are communities, villages, *rancherias* and tribes, etc," and expressed the need of giving explicit recognition in the Constitution to the corporate property rights of the villages, as in colonial times.[50] The framers of the Constitution approached the question of communal property by recognizing its legitimacy and the legal capacity of the communities to own such property corporately, although they left open the door for the future division of communal property among the members of these communities: [51]

> Properties held in common by co-owners, hamlets situated on private property, villages, communities, tribes, and other settlements which in fact or by law conserve their communal character, shall have legal capacity to enjoy in common the lands, woods, and waters which belong to them, or which may have been or shall be restored to them

> in accordance with the law of January 6, 1915, until such time as the manner of making the division of the lands only (as distinct from the woods and waters) shall be determined by law.

Restitution of the communal lands lost by the village communities was the next task undertaken by the authors of article 27. By means of paragraph VII- 9 of Article 27 they went a step beyond the 1915 decree, and declared null and void all the transactions since June 25, 1856—even when carried out legally—that had deprived these communities of their lands, requiring that restitution take place.[52] An exception was made for small properties below fifty hectares legally sold after 1856:[53]

> All proceedings, findings, resolutions and operations of demarcation, concession, composition, judgment, transaction, alienation, or auction which may have deprived properties held in common by coowners, hamlets situated on private property, villages, communities, tribes and other settlements which exist since since the law of June 25, 1856, of the whole or a part of their lands, are declared null and void; all the findings, resolutions and operations which take place in the future and produce the same effects shall likewise be null and void. Consequently, all the lands, woods, and waters of which the above-mentioned communities have been deprived, shall be restored to them in accordance with the decree of January 6, 1915, which shall remain in effect as a Constitutional law.

The new Constitution thus reestablished explicitly the capacity of the villages to hold land property that the 1856 Constitution had taken away from them, confirmed the legitimacy of their property rights over land still being held by them since colonial days, and made it possible for them to reclaim ownership of the lands lost since 1856. The lands that were recovered by the villages through restitution under these clauses came to be known over time as *tierras comunales*, and its owners as *comuneros*. The villages had to produce satisfactory evidence that the lands they claimed by way of restitution were actually in their possession as communal property before or after 1856; this was not always the case, and many claims of restitution were actually shown to have no basis at all.

While restitution of the lands lost to the villages gave the Constitutionalists the initial motive and moral justification for their legislative actions, restitution was only a minor part of what they saw as the more general problem of Mexican agricultural development: the concentration of property in a relatively small number of large estates. The division of the large *haciendas* among a great number of owners was for them the really significant objective of general land reform.

By introducing the principle that all property is ultimately derived from the sovereign, the government was given the power of eminent domain, whereby it could "regulate the status of all property with payment of the respective compensation." This would "permit the government once and for all, to solve with ease the most difficult part of all the property matters involving the agrarian problem, which is the division of the large estates, without loss to the owners." They specifically proposed breaking up "the latifundia through expropriation, but with respect for the rights of the owners." This was formulated in the Constitution in the following terms:

> The nation shall at all times have the right to impose on private property such limitations as the public interest may demand..... For this purpose necessary measures shall be taken to divide up the large landed estates; to develop small landed holdings; to create new centers of rural population with the lands and waters that are indispensable to them; .....villages, hamlets situated on private property, and communities which lack land and water, or do not have them in sufficient quantities for the needs of their inhabitants, shall have the right to be provided with them from adjoining properties, always having due regard for small landed holdings. Therefore, all grants of lands made up to the present time under the decree of January 6, 1915, are confirmed. Private property necessary for the achievement of the above-expressed objectives shall be considered as taken for public utility.[54]

They thus went beyond restitution and made possible new endowments or grants of land, *dotaciones*, to peasant communities which could not claim restitution. The peasants were entitled to organize themselves to petition for land, which would be granted to them by dividing the large estates:

> In case that the adjudication of lands that may have been requested by any of the above-mentioned communities by way of restitution be not legal in the terms of the said decree, those lands shall be given to them by way of grant, and they shall in no event fail to receive such as they may need.[55]

The new land grants that were made possible under this clause came to be known later as *tierras ejidales* and its owners as *ejidatarios* as distinguished from *tierras comunales*, whose rights are derived from pre-1856 possession. The term *comunal* land became synonymous with restitution in the 1934 Agrarian Code, while *ejidal* land became linked to endowment or *dotacion*—a distinction which has been kept until today. The creation of new *ejidos* became the main activity, by far, of the agrarian reform programs. Around 85% of the land distributed in the seventy five years since 1917 belongs in the *ejido* category of new endowments while only around 15% was obtained through restitution.

The assembly tought of empowering the states to legislate the allowable maximum size of the *haciendas* in accordance to their local needs, making it mandatory upon their owners to sell any remainder above the limit. Only where the owners refused to do so would the state governments intervene and expropriate the land in order to achieve these goals. Thus a double strategy was established by the framers of the Queretaro Constitution to eventually reach the ideal of a more egalitarian rural society: one based on the sale and subdivision of the *haciendas* into smaller private landholdings, and the other based on expropriation and *ejido* restitution and endowment.

Whereas the Constitution of 1856 required that where expropriation was due, it could only take place "prior compensation", the framers of article 27 simply stated " by means of compensation."[56] Aware that compensation to the estate owners for land taken for land endowments implied a heavy debt burden for the country, the committee report to the assembly assumed that expropriated land "would be paid for by the same ones who acquire it, thus reducing the role of the State to that of a simple guarantor." This early assumption that the grantees should themselves pay for the land endowments was in contradiction to the decree of 1915—which was made part of the Constitution—that land should be provided at government expense. The committee's recommendation, nevertheless, was applied during Carranza's lifetime, by making each *ejidatario* sign a bond commiting him to reimburse the compensation amount paid out to the former owner.[57] That practice, however, was given up completely as soon as Obreg6n became President.

Concerned with the possibility of a future repetition of the process that had led in the past to the loss of the communal lands and the formation of latifundia the framers of the new Constitution imposed heavy restrictions on the property rights of the *ejido* communities, explicitly outlawing in the text the alienation of the lands obtained by restitution or endowment. The historical precedent of the Díaz period, in which the villages had lost their land through market pressure and illegal usurpations, was in the minds of the framers of the 1917 Constitution. They sought to close any possible avenue that might lead to such an outcome by explicitly outlawing in the same text the alienation of the lands obtained by restitution: "the rights to the same lands shall be inalienable."

The door was opened in the Constitutional text for the future internal division of the agricultural land of the *ejido* communities among its members "until such time as the manner of making the division of the lands only, shall be determined by law" (Art. 27-VI). The text, however, envisioned the individual property rights resulting from future division to be equally inalienable:

> "Only the members of the community shall have the right to the lands to be divided, and the rights to the same lands shall be inalienable while they remain undivided, as well as those held in ownership when the division has been made."

The division of the communal agricultural holdings among its members was also accepted by the radical wing of the Constitutional assembly—the so called Jacobins—including Francisco J. Mugica, one of the leading figures of the convention, who only wanted to make sure that division would not apply to the woods and waters "which should be enjoyed collectively, not being divided under any circumstance."[58]

The 1917 Constitution thus created a form of **corporate private ownership** held by peasant communities and not by the state as in the Soviet Union. Using as a model the old communal form of property, the drafters of the Constitution granted legal personality to the communities to recover, hold, and defend property in common, and to enjoy in common lands, woods, and waters "which belong to them, or which may have been or shall be restored to them,"[59] and that no private interest could alienate from them. It made these property rights inalienable and untransferable, as one of the means of preventing the creation of latifundia and of protecting the village communities from losing the source of their livelihood. The property rights of the *ejidos* were even made *imprescriptibles* which meant that they would not prescribe through possession by any other occupant no matter how many years had transpired.

Although the historical village community and the modern post-revolutionary communities and *ejidos* have much in common, they are far from identical. An important difference concerns the much greater economic and political autonomy that was enjoyed by the historical communities as a result of the *Tierras de Propios y Arbitrios* that they held to provide for their local collective needs. The revolutionary *ejido* communities were not granted such resources to sustain their local governments, which as a result have remained dependent on the favors of the federal government that could thus demand political allegiance and commercial production in exchange.[60]

Whatever ideological differences divided the drafters of the 1917 Constitution, it is clear from the debates, committee reports, and the

Constitutional text, that there was full consensus on the fundamental legal nature of communal ownership rights as a form of property different from state or government ownership. For both the radicals and the moderates, the *ejido* belonged to the peasant communities as corporate private property, and not to the state or the government. Its non-marketability was not a result of the government retaining ownership of the land as in the case of the subsurface resources, but was a restrictive safeguard to prevent its loss by the peasants and the re-creation of the latifundia. Nowhere is there even a hint of the *ejidos* being a form of usufruct of national lands, a legal form derived from Roman law otherwise very well known in Mexican civil law.

THE IMPLEMENTATION OF THE AGRARIAN REFORM

The Constitutional Assembly in Queretaro marks only the beginning of the long era of agrarian reform in Mexico which continues to this day. With the end of the Constitutional assembly began the difficult task of of translating into specific legislation the general principles laid down in Queretaro and of implementing the agrarian reform. Although these various legislative efforts reflect the changing moods of the state with respect to the *ejido*, and the efforts to steer it into different directions by the various governments according to their agrarian philosophy and the internal and external circumstances of the moment, the basic nature of the *ejido* as laid down in Queretaro could not be so easily changed.

Thus, although one can certainly recognize the different currents trying to influence *ejido* policy, the basic principle that the *ejido* is the corporate private property of the community, not state property given in usufruct, has survived. A brief survey of these changing agrarian policies up to the recent past will be attempted within the general context of the general economic and political environment. Four broad historical periods[61] can be recognized in this respect: from 1915 to the Depression and the end of the Calles era; from 1934 to 1940, during which President Lázaro Cárdenas injected renewed vigor into Mexican nationalism and the agrarian reform; from World War II until 1970 and the end of the economic boom-period; from 1970 to 1982, which was marked by recession and statist expansion.

The First Period: (1915-1934)

The early actors of land reform during the twenties understood land reform as a means to an end: the creation of a prosperous system of small commercial farms. President Calles, a dominant figure of this period, made clear in his introduction to the Irrigation Law that the goal of land reform was "to create a class of small landowners and to dissolve the communal form of land tenure." Fully aware of the need to produce surplus food to feed the cities, Carranza, Obregón and Calles were nevertheless reluctant to divide the large estates of the northern and central states, where the owners were still powerful.

Government turned its attention in the second half of this period to the development of commercial agriculture in the North and Northwest, which became the main beneficiary of government investment. There, the government undertook the first irrigation projects and rural roads programs which were continued and expanded in the later decades. Convinced of the superiority of commercial farming over subsistence agriculture as practiced by the traditional *ejidos*, Calles transfered these rich irrigated lands to private farmers who applied more advanced technology, and became highly successful in developing large-scale commercial agriculture, responsive to the national and export markets.[62]

Yet the revolutionary leaders needed the subsistence mini-plot *ejidos* for the dissatisfied peasants of the South who followed Zapata in his struggle for "land and liberty," or, as formulated by President Calles, "as a school to train peasants in the management of farms, before they could become full-fledged farmers." The leaders in this period shared the Cabrera view that the *ejido* should be seen as an income supplement to the salary of the peasants working on the *haciendas*, while the *hacienda* system was still in place. In the densely populated southern states of Morelos, Oaxaca, Guerrero, Puebla, Veracruz, and the Distrito Federal, where peasant claims were far more demanding and the estate owners weaker, the country's leaders implemented the *minifundio* which turned out to be anything but transitory, and became instead a permanent component of the agrarian reform.

The individual *ejido* parcels in the central states were not larger than three or four hectares, often on land of marginal quality, lacking water and, with little or no credit, on which the peasants grew maize for their own consumption using primitive technology. "Essentially" says Clarence Senior, "it was a subsistence farm with the peasant barely able to keep body and soul together, if the neighboring *hacienda* did not give him day labor. But generally the Indian peasant was satisfied. His hunger of land had been allayed." Most of the land distributed was granted to the peasants under the *ejido* endowment clause and not on claims of restitution (see table 13). The minimal importance of restitution becomes clearer when judged from the standpoint of the number of beneficiaries, 98.7% of whom were *ejidatarios* and only 1.3% of whom were *comuneros*.

The first period basically recognized only one type of *ejido*, the parcelized *ejido,* in which the agricultural land of the community was divided into separate parcels among the individual *ejidatarios.*[63] The legislation of this period which includes the *Ley de Ejidos* of 1920 and the *Reglamento Agrario* of 1922 by President Alvaro Obregon defined the minimal size of the individual miniparcels as a function of the availability of water and the quality of the soil. These minimums varied from 2 to 4 hectares for irrigated land, and from 4 to 5 hectares for better seasonal land. These minimums varied over time, reflecting the changing policies towards the *ejido*, although the general tendency clearly points to increased size. This analysis is summarized in table 13.

Although the Law of Parcelized *Ejido* Property of December 1925,[64] required that parcelization be carried out within a maximum period of four months after an endowment to a group had taken place, the question of the final parcelization and assignment of tenure titles to individual plots to each *ejidatario* was left pending. It was not until 1940 that the question of individual titles in the mini-plot *ejidos* began to take shape in the agrarian legislation. In any case one would not find a departure in this period from the principles laid down in Queretaro concerning the basic nature of the *ejido* property rights.

The basic characteristics of internal governance of the *ejido* were established in the first period. Authority to manage the exploitation of the lands of each *ejido* was vested in 1917 in an elected administrative committee of three members.[65] The Law of 1925 added the requirement that an annual assembly be held in the *ejido* to elect the committee, which was placed under the surveillance of a delegate or representative of the Comisi6n Nacional Agraria. An amendmentto that Law in 1929 established the requirement of appointing a vigilance committee at the side of the administrative committee.

The administrative committee was put in charge of distributing the parcels among the *ejidatarios*, which it often carried out in an arbitrary fashion, using its power as an instrument of pressure over the *ejidatarios* for political purposes. The continuous history of abuse of power by the *ejido* authorities and the internal conflicts that it generated, which have plagued the *ejidos* over the years, have their beginning in this early period. The *ejidos* were controlled very early on by politicians who used them "as salable commodities in the political market."[66]

Table 13. *Comunal* Land Restitutions and *Ejido* Land Grants by Presidential Periods in Mexico (1915-1992)

Period		(a) Restitutions Comuneros (sq km)	no.	(b) Land Grants Ejidatarios (sq km)	no.	(c) Total Beneficiaries (sq km)	no.
Period 1							
Carranza	(1915-20)	829	3,633	2,990	73,570	3,819	77,203
Obregón	(1921-24)	2,582	3,126	14,725	161,002	17,307	164,128
Cálles	(1925-28)	3,4045	3,969	28,458	298,570	31,863	302,539
P. Gil	(1929-30)	23	66	24,362	187,203	24,385	187,269
O. Rubio	(1931-32)	3,155	1,160	9,103	56,834	12,258	57,994
Rodriguez	(1933-34)	90	131	20,513	158,252	20,602	158,393
Subtotal	(1915-34)	10,083	12,085	100,151	935,431	110,234	947,526
(%)—>		(1.0)	[0.3]	(10.2)	[28.7]	(11.2)	[29.0]
Period 2							
Cárdenas	(1934-40)	389	1,679	201,070	763,209	201,459	764,888
(%)—>		(0.1)	[0.0]	(20.4)	[23.4]	(20.5)	[23.4]
Period 3							
Avila C.	(1941-46)	6,635	10,834	53,069	112,107	59,704	122,941
Alemán	(1947-52)	12,291	17,571	42,105	91,054	54,395	108,625
Ruíz C.	(1953-58)	22,079	30,593	35,639	195,699	57,717	226,292
L. Matéos	(1959-64)	11,579	28,878	81,503	260,478	93,082	289,356
D. Ordáz	(1965-70)	48,611	107,748	181,945	266,772	230,556	374,520
Subtotal	(1941-70)	101,194	195,624	394,261	926,110	495,454	1,121,834
(%)—>		(10.3)	[6.0]	(40.1)	[28.4]	(50.4)	[34.4]
Period 4							
Echeverria	(1971-76)	33,828	89,732	86,342	129,186	120,171	218,918
L. Portillo	(1977-81)	11,553	55,537	44,536	155,057	56,089	210,594
Subtotal	(1940-70)	45,381	145,269	130,878	284,243	176,260	429,512
(%)—>		(4.6)	[4.4]	(13.3)	[8.7]	(17.9)	[13.2]
Total	(1915-81)	157,046	354,697	826,360	2,908,963	983,406	3,263,660
(%)—>		(16.0)	[10.9]	(84.0)	[89.1]	(100.0)	[100.0]
la Madrid	(1982-88)					50,812	
Salinas	(1988-84)					7,015	

Source: Zaragoza J.L. & Macías R. (1980), CENIA (1981)

Table 14. Number of *Ejidos* in México by Total Size (1970)

Size (sq. km.)	No.	%
Less than 0.5	184	0.8
0.5 to 1.0	388	1.7
1.0 to 2.0	996	4.4
2.0 to 5.0	3,912	17.3
5.0 to 10.0	4,856	21.4
10.0 to 50.0	9,420	41.5
More than 50.0	2,935	12.9
Average 30.7 sq. km.	22,692	100.0

Source: V Censo Agricola y Ganadero, 1970. Avellaneda, 1976

The basis for a credit system to assist the *ejidos* was laid in the first period by Calles. He created a unified system of agricultural *ejido* banks (Bancos Agrícolas Ejidales) to assist the *ejidos* with credit, agricultural training, and cooperative organization on March 16, 1926. Nine regional banks were establishedin various states of the Republic between 1926 and 1930 to provide credit to those *ejidatarios* "who would organize themselves in local agricultural cooperatives authorized by the Ministry of Agriculture."[67] These specialized *ejido* banks were replaced in 1931 with general agricultural banks that were authorized to operate with small private landowners as well as with *ejidatarios*. Their credit operations were also restricted to *ejido* cooperatives that included the majority of the *ejido* members, excluding individual *ejidatarios* for reasons of financial security, which unintentionally resulted in strengthening the cooperative character of the *ejido*.

A total of 6,641 *ejidos* was created in the first period, half of which were located in the central states (48%), with an average size of 1660 hectares/*ejido* and an average membership of 142 peasants/*ejido*.[68] In total, the governments of the first period distributed 110,234 square kilometers to 947,526 peasants, the equivalent of 11.6 has/*ejidatario*.[69] The program benefitted 7.9% of the total rural population of 12 million people, but only increased their share in the cropland of the country from 10% to 13% from 1910 to 1930 (see table 12).

The *hacienda* system was not destroyed in the twenties, although it was badly enough damaged to induce a transfer of capital and labor from the rural to the urban areas, particularly to Mexico City. It was not until the Great Depression in 1930 that the weakened *haciendas* received their *coup de grace,* when the agricultural export markets, on which much of their commercial farming was based, lost their economic base. The collapse of the foreign markets also hit the other sectors of the economy and induced a recession in the traditional Mexican exports of minerals and petroleum. Mexico's economy, which had always swung with the ups and downs of the world market, was deeply affected, as its real GNP had fallen below that of 1910.

The Depression changed the direction of the political winds in Mexico. Calles, who had thought of slowing down or bringing to an end the establishment of more *ejidos,* lost ground to the radical *agraristas* in the revolutionary coalition. The *agraristas* perceived this historical moment as a unique opportunity for dismantling the *haciendas,* intensifying land distribution and creating more *ejidos.* With the nomination of Lázaro Cárdenas in Queretaro as the presidential candidate of the official party in 1933 came an end to the hegemony of the Northern revolutionary leadership. A new period began in which the Northern view of land reform was substituted for the next six years by a radical blend of state corporativism and some elements of the Southern ideology.

The Second Period: The Cárdenas Era (1934-1940)

The second period of the agrarian reform (1934-40) is synonymous with Lázaro Cárdenas. He made land reform, the centerpiece of his government program. His militant agrarian policy not only accelerated the pace of *hacienda* land distribution and the creation of new *ejidos,* but also radically reinterpreted the agrarian reform program. The *ejido* was raised from a supplementary source of income for the field laborer to the rank of one of the main pillars of agricultural production, on whose shoulders was placed the responsibility of feeding the country, which the *haciendas* previously fulfilled. In his famous address in Torreón on November 30, 1936 Cárdenas made clear his program for the *ejidos* in the north:

> There could have been some who in the early days of the Revolution considered the *ejido* as a mere supplement of the laborers salary, insufficient to guarantee him economical independence, which is the basis of all freedoms. That carries no weight in the present decisions of the authorities....
>
> The *ejido* institution today has a double responsibility: as a social system, inasmuch as it frees the field-laborer from the exploitation of which he was the object in both the feudal and the individual [land

> tenure] systems; and as a system of agricultural production, inasmuch as the responsibility of feeding the country weighs upon the *ejido*.[70]

Cárdenas moved the *ejido* program to the more prosperous newly irrigated districts of the northwest, where he intended to carry out his plans of combining *ejido* organization with efficient commercial farming. There, where Calles had begun to build new dams and irrigate rich lands, Cárdenas expropriated prosperous commercial plantations as total units, and handed them over to their laborers, who took them over as collective *ejido* property. These *ejidos* were assigned the economic role of producing the basic food staples to feed the whole country. Through government credit and technical assistance, they were brought into a government-directed system of commercial farming.

Cárdenas introduced an important innovation in the *ejido* system which was intimately linked with his plans for a new economic role for the *ejidos* as commercial food producers, creating at the side of the parcelized *ejidos* the collective *ejidos*. These collective *ejidos* were organized as cooperative farms where cooperative work was added to collective ownership in what was defined then by a group of socialist economists as "*haciendas* without *hacendados*." Flourishing *haciendas* were expropriated and transformed into collective *ejidos*. The collective *ejido* was seen by Cárdenas as the only formula for preventing these flourishing commercial agriculture units from reverting to subsistence agriculture. Around 900 collective *ejidos* were created between 1936 and 1938.[71] The first collective *ejidos* were organized in the Laguna Region, and after that in the Mexicali region of Baja California, the Yaqui Valley of Sonora, Los Mochis in Sinaloa, New Italy and New Lombardy in Michoacán, and in the peninsula of Yucatán at the southern tip of the country.

To push forward agrarian reform, Cárdenas made the necessary legislative changes to allow a substantial increase in *hacienda* expropriations on the one hand and potential *ejidatarios* on the other. To facilitate these objectives the outgoing government of President Abelardo Rodríguez approved amended article 27 of the Queretaro Constitution and unified all the legislation pertaining to agriculture in the Agrarian Code of July 18, 1934.[72] The code eliminated most exemptions which previously sheltered many private properties above the maximum size (150 hectares for irrigated land and to 300 hectares for seasonal land) from expropriation. The 1937 amendment by President Cardenas also decreed that all the separate properties of a single estate-owner should be considered as one plot to determine size.[73] Thus was a far greater amount of *hacienda* land made available for expropriation and distribution during the 6 years of the second period, than in the total previous 19 years. A total of 201,459 square kilometers were distributed to 764,888 peasants (table 13).

Cárdenas also eliminated the previous restrictions on aspiring *ejidatarios* and included among them the *acasillados*, the *peones* not living in the free villages but on the *haciendas*, who until then were not entitled to request or receive *ejido* plots.[74] One of the results of granting land to the *acasillados*, was a major reduction in the size of the *hacienda* labor force from 3 million in 1910 to 800,000 in 1940.[75] The average amount of land allotted per *ejidatario* grew nationally from 11.6 hectares to 26.3 hectares (table 13). However, in the densely populated central and southern regions, Cárdenas continued the old policy of his predecessors of creating subsistence *ejidos*. The latest agrarian Code of 1940[76] issued by Cardenas still left the minimal size of the individual agricultural parcels as 4 hectares for irrigated land and 8 hectares for seasonal land.

For all his strong inclination toward the collective *ejido* which he saw as more efficient in a economic sense, Cardenas neither attempted to eliminate nor restrict the parcelized *ejido* which, by far, remained the dominant form, as pointed out even by the collectivist critics of this aspect of his policy.[77] The collective *ejidos* were allowed to return to a parcelized organization, if that was the will of the community—which almost all of them actually did at the end of the Cárdenas period. He clearly expressed this in his well-known address in Torreón previously mentioned: "Collective organization of the *ejido* does not mean that there is any intention of excluding different organization forms: far from it, where individual operation is economical it is implemented and stimulated."[78] Thus in the Agrarian Code of September 23, 1940, approved by Cárdenas before the end of his presidential term, he laid the legal basis for what was later given full form in the 1942 Code of granting the *ejidatarios* restricted individual tenure titles to their agricultural parcels.

Being aware of the crucial role of credit to the success of his *ejido* program, Cárdenas adopted once again the concept of a specialized bank for the *ejido* system, in charge not only of financial matters but of the economic organization of the *ejidos*. He used credit as a lever to strengthen the collective character of the *ejidos*, disqualifying credit transactions with individuals by the bank, and only granting access to credit societies containing a minimum of 51% of the parcel holding members,[79] a requirement which only 40% of the *ejidos* actually satisfied.

The leadership of the *ejidos* was coopted and turned into an instrument for controlling political demand-making and for mobilizing regime support in the rural areas. Through the *ejidos* and their leadership the government increased the state's influence in the countryside and its control over the organization of peasant groups, which were incorporated into the regime by way of the dominant party structure. The National Confederation of Peasants, the CNC, was created and subsequently incorporated as one of the three corporate pillars of the dominant political party in this period. *Ejido* elections became prey to constant interference.[80]

The *ejidos* were made highly dependent on the favors of the regime. A perpetual clientistic relationship was established between the government and the *ejido*. "Cárdenas and those that followed"—says Tutino—"used their powers in the *ejido* communities to demand political allegiance as well as commercial production...the *ejidatarios* received more land than autonomy."[81] On their own they were not granted the resources to play the role assigned to them as one of the main pillars of the economy. They became permanently tied to the regime by the credits, markets, extension services, infrastructure, and inputs that the government could provide—a dependent clientele of the state in every respect. Without these resources most were barely able to subsist.[82]

While it is certainly true that the *ejido* became highly dependent on the state in the Cárdenas period, President Cárdenas did not change the basic nature of the *ejido* property rights. The concept of the *ejido* as a form of state property given in usufruct to *ejido* communities was never suggested, much less adopted, by Cárdenas. One does not find in the Cárdenas collective *ejido*, or in the agrarian legislation adopted during his administration, any support for such a view. The *ejidos* remained the property of the peasant communities and not state property. The collective *ejidos* which he created operated within the same basic framework of the communal property rights applied to the parcelized *ejidos*.

The end of this period witnessed a split in the Cárdenas camp and polarization between the left and right-wing followers. While the left-wing led by General Múgica, one of the participants of the 1917 Constitutional assembly, was in favor of continuing unabated the land distribution program, those to the right reacted against the *ejido* and in favor of the small private property. General Almazán, a right-wing member of the Cárdenas cabinet who launched his dissident candidacy for president in 1939, attacked what he claimed to be the revival of serfdom. He proposed that *ejido* rosters be purged of undeserving members with the purpose of dividing the *ejidos* into average parcels of twenty hectares that could provide a decent living to the legitimate *ejidatarios*. The *ejidatarios* would then be granted fee simple private titles to become small private owners, in order "to free them from political manipulation and misery." Neither Múgica nor Almazán reached the presidency, which went to General Manuel Avila Cámacho, a middle roader who opened the third period of agrarian reform.[83]

World War II and the Post-War Period (1940-70)

The beginning of the third period of agrarian reform is marked by the start of World War II and the end of the Cárdenas presidency. Just as the Great Depression led to the implementation of radical agrarian reform, World War II brought about a reorientation of that same reform. The developments in the Mexican economy that took place as a result of the new international situation enhanced the economic role of the commercial private farms and de-emphasized the role of the *ejidos*—most of which remained subsistence oriented.[84] A partial return to some of the ideological premises that dominated the first period took place, which explains the "counter-reform" label that was given to this period by its critics from the radical camp.

World War II reversed the devastating effect of the Great Depression on Mexican commercial export agriculture, producing a rebirth of agricultural exports. War shortages in the United States raised commodity prices and assured Mexican producers a high rate of return on exported agricultural goods. Export-oriented commercial agriculture, a fundamental component of the old Porfirian economic model, was once again called to play a primary role in the war years and the 2 decades after, as the main source of foreign exchange. The agricultural sector became the "engine of growth" for the national economy that provided the foreign currency to pay for import substitution industrialization, which the then stagnating primary industries of petroleum and mining were unable to generate. Increased agricultural production for export became a major government objective.

Ways were sought to raise agricultural production by imbuing a sense of security and confidence in the private agricultural producers. A new Agrarian Code was adopted in 1942 by President Manuel Avila Camacho,[85] granting certificates of immunity against further expropriation to private landholdings of up to 300 hectares producing key export commodities. Private owners protected by such certificates of immunity were further granted, in the 1947 constitutional amendment to Art. 27 by President Miguel Alemán, the *derecho de amparo*, the right to appeal and ask for an injunction against government expropriatory action where they were actually engaged in agricultural production.[86]

The conciliatory economic policies followed in this period brought about remarkable growth of the Mexican economy, as reflected by the basic indicators of that period. The agricultural policies were highly successful from the standpoint of increasing production and productivity until 1965. Growth of agricultural production reached a yearly rate of 5.5% during the 1940-55 period and 4.1% during the next decade.

The North and Northwest, containing 75% of the new farm lands created by massive irrigation and rural road projects, became the center of Mexican commercial agriculture. Large commercial farms, partially or totally mechanized, linked to specific export crops were organized. The land rights in the irrigated lands were almost equally divided between private owners and *ejidos,* which could thus benefit from the new favorable opportunities. However, the resource-poor *ejidos* found themselves less capable of reaping the fruits of these opportunities which were much dependent on high-value production inputs. Nevertheless, they kept the pace set by the private sector and produced a significant share of total agricultural output, market farm products, and export crops.[87]

Important changes of policy with regard to the *ejido* institution took place within this new context. A greater concern for the role of the individual *ejidatario* characterizes this period. To increase production in the *ejidos* and induce land shepherding and investment by the *ejidatarios* in their parcels, individual tenure security was seen to be a key factor. Legislative steps were taken to grant each *ejidatario* individual tenure rights over a specific parcel within the *ejido* which he could pass on as an inheritance to his nearest kin. The Agrarian Code of 1942 established individual tenure titles to "render the security of possession to each *ejidatario.* " Two types of possession titles were established in the 1942 Code, *Titulos de Usufructo Parcelario,* Titles of Parcel Usufruct, for *ejido* parcels fulfilling the minimal area requirements established in the agrarian legislation, and *Certificados de Derechos Agrarios,* Certificates of Agrarian Rights, for those holding parcels below those minimum limits. The introduction of these titles is probably the most important single development in the post-revolutionary period, as far as *ejido* property rights is concerned, within the basic constitutional concept that the *ejido* is the property of the peasant communities, not of the state.[88]

The individual *ejido* titles or certificates recognized in 1942 were not fee simple property titles. They did not in any way relax the fundamental non-marketability restrictions applying to *ejido* property, nor did they sever the links of the individual *ejidatario* to the *ejido* community. They were intended to bring an end to the system of rotation of parcels which was seen as a source of inner manipulation within the *ejidos* and an obstacle to land shepherding by the ejidatarios.[89] The term usufruct was introduced for the first time in this period in the denomination of the individual tenure titles (*Titulos de Usufructo Parcelario*), not to suggest in any way government property of the land but to impress upon the minds of the *ejidatarios* that their parcels were not fee simple property titles.[90]

The almost total disappearance of the collective *ejido* characterizes the third period. Not only did the creation of new collective *ejidos* cease almost completely, with the exception of a few cattle raising *ejidos* in the Northwest in the late sixties, but most of the collective *ejidos* that were established during the Cárdenas period reverted to parcelized organization. Although the collective *ejidos* had generally shown themselves to be economically more viable than the parcelized *ejidos*, according to some analysts,[91] they were not yet able to stand on their own feet when they lost much of the official support they enjoyed in the Cárdenas days. Torn by inner dissension, most of them disappeared and reverted to the parcelized type.

The official *ejido* credit policy also underwent changes to promote greater individual initiative and responsibility. The Agricultural Credit Law adopted in 1942 and its reforms of 1945, 1946, and 1947, and the later Credit Law of December 30, 1955, represent a significant departure from the Cárdenas approach. The requirement that the majority of the *ejidatarios* had to participate in a credit society in order to qualify for credit was eliminated. Multiple credit societies of ten members could be organized within an *ejido*. Even smaller groups of three or more *ejidatarios* sharing common responsibility were admitted for the first time, and credit operations with individual *ejidatarios* were given recognition.

The agrarian reform in the first eighteen years of the third period during the presidencies of Avila Camacho (1941-46), Miguél Alemán (1947-52) and Adolfo Ruíz Cortínes (1953-58) returned to the slower pace maintained before Cárdenas became President. These three presidents each distributed an average of 57,272 square kilometers of land to a total of 457,858 *ejidatarios*, which come out to an average of 37.5 hectares/*ejidatario*. Land distribution, however, started to gain momentum with President López Matéos (1958-64) who distributed 93,082 square kilometers, reaching its climax during the regime of President Díaz Ordaz (1964-70), who distributed more land than Cárdenas, 230,556 square kilometers, to 374,520 peasants.

In the 30 years of the third period as a whole (1940-70), a total of 495,454 square kilometers were distributed among 1,121,834 *ejidatarios* coming out to an average of 44.2 hectares/*ejidatario* (table 12).[92] This area represents half of the total area distributed between 1915 and 1980 and a third of the total number of beneficiaries, which hardly justifies the "counter-reform" label given to this period. At the end of this period, there were 21,475 *ejidos* and 1,206 *comunidades* in the country, with a total of 2,182,486 members holding around 47% (807,147 square kilometers) of the total cropland of the country.[93]

The literature is highly critical of some features of the *ejido* in this period. Although the principle of individual tenure rights remained a basic tenet of the agrarian legislation, only a minority of *ejidatarios* actually received these titles over time. The *ejidatarios* had to struggle to obtain these titles with the agrarian bureaucracy, at enormous cost to themselves in time, energy, and money.[94] The lack of title allowed the *ejido* establishment to maintain the system of parcel rotation and manipulation within the *ejidos*, penalizing those who produced more and discouraging any kind of soil shepherding.[95]

The *ejido* was not allowed to develop internal democracy. Imposition of the *ejido* leadership from above which characterized the *ejido* from its earliest history did continue unabated. Being charged with the duty of supervising elections, the higher agrarian bureaucracy became an instrument to control *ejido* elections to the benefit of outside economic and political interests. In many instances *ejido* elections were simply omitted and the appointments were made from the top. The bureaucracy of the *ejido* banks serving the *ejidos* became the constant target of accusations of corruption and fraud (Hansen, 1971).[96]

A slowdown in agricultural growth was suddenly experienced in 1964. Agriculture, the main exporting sector of the economy, began decreasing its total exportable output after 1965, reducing the inflow of foreign reserves with which to finance industrial imports. The agricultural growth rate fell to 2.3% in the 1964-69 period, substantially below the annual demographic growth rate of 3.4%. This created the background for the political crisis of 1968, which once again changed the direction of modern Mexican history. The country's agricultural self-sufficiency ceased for the first time in modern times in 1970, and a period of increasing food imports and lending began.[97]

The question of which factors were responsible for bringing an end to the so called Mexican economic miracle is still open to debate. Was the slowdown related to the unprecedented magnitude of the land distribution program of President Díaz Ordaz, which exceeded even that of Cárdenas? Was it the result of a process of decapitalization of the agricultural sector through low prices and wages, as argued by others?[98] Or was it externally induced by international recession? Whatever the answer, it is now clear that the third period came to an end when the winning political faction of the 1968 watershed-year seized the opportunity to change the course of the economic policies the country had followed since 1940.

The Fourth Period of the Agrarian Reform (1970-82)

The fourth period of the agrarian reform begins with the presidency of Luís Echeverria Alvarez in 1970. A hindsight evaluation of the Echeverria spending pattern and policies has leds analysts to the conclusion that his main goal was the accumulation of greater power in the hands of government through the economy. His regime, as now understood, intended to bring about state domination of the national economy. As viewed by these analysts the expansion of the role of the state in the economy became a more important goal than the expansion of the economy itself.[99]

Populism, as a way of winning political support from the corporate popular base, was another major component element of his government policy. Rather than choose between industrialization and social welfare, spending was increased in both areas, to bring into a political coalition the advocates of public-sector industrialization with those emphasizing social welfare. As a result government expenditures rose in relative and absolute terms between 1970 and 1975, from 13.6% ($US 3 billion) of the GNP to 40% of the GNP ($US 12 billion).[100]

The statist-populist policy was carried out without tax reform and was financed using credit from internal and international sources. Internal credits to the government from the Central Bank tripled from 1971 to 1976, while credit to the private sector remained constant. The foreign debt multiplied from $US 3 billion to $US 17.9 billion. The number of public or mixed enterprises climbed from 400 in the early seventies to 845 in 1976, and the number of employees in the different federal goverment bodies reached two million people.[101]

The government faced the problem of maintaining agricultural production for export to generate foreign currency vital to the Mexican economy. The three independent agricultural bank systems, the Banco Ejidal, Banco Agrícola, and Banco Agropecuario, were consolidated into one unified bank, the Banco Nacional de Credito Rural (July 5, 1975), and unprecedented sums of money were made available to the bank to back the government's agricultural plans.[102] Public sector investment in agriculture was greatly expanded in absolute and relative terms from 13.2% of the federal budget in 1970, to 20% in 1975.[103]

In response to the agricultural crisis, renewed emphasis was given to the *ejido* institution in a political move which emulated the Cárdenas turn during the Depression. The Departamento Agrario DAAC was raised to ministerial rank. The Zapata and Cárdenas rhetoric was revived in an attempt to replicate in the countryside the peasant enthusiasm of the Cárdenas era. The president promised to do in five years what would otherwise take fifty years to accomplish.[104] A new Agrarian Reform Law was approved in 1971 superseding the 1942 code.[105] The legal minimum size of the individual *ejido* parcel was raised in the law to 10 hectares for irrigated land, and to 20 hectares for seasonal land. A total area of 120,171 sq. km. was distributed to 218,918 *ejidatarios* during the Echeverria presidency, which places this period in third place after Díaz Ordaz and Cárdenas as to the amount of land distributed.[106]

The collective *ejido* was once again given high priority in government policy: "The *minifundio* as a form of production, contradicts the very nature of the *ejido* system. Where possible, and wanted by the peasants, we will create by parcel merger and cooperative effort veritable *ejido* enterprises," promised the new President in his inaugural address on December 1, 1970. The agrarian bureaucracy announced plans for the organization of 7000 collective *ejidos* .[107] An incentive to the parcelized *ejidos* to turn to collective organization was offered in the new Law of Rural Credit by establishing preferential interest rates for the collective *ejidos.*[108] Reversal from the collective form of organization to the parcelized form in any *ejido* was made more difficult and dependent on approval of the President himself, requiring a technical study and a two thirds approval vote by the *ejidatarios.*

Notwithstanding the Ministry's efforts to educate the *ejidatarios* to the virtues of collectivization, the campaign found no resonance among the peasants. The enthusiasm that existed in the days of Cárdenas for the collective *ejido* had vanished among the *ejidatarios*, who did not respond to the overtures of renewed collectivism. The collectivization program, to the contrary, fueled rumors and increased tensions in the countryside, while official corruption added to the woes of the program.[109]

During this period another important development in *ejido* development policy took place which sought to link the *ejido* to non-agricultural economic activities. The President introduced his development scheme for the *ejidos*, stating: "The *ejidos* should not only dedicate themselves to agriculture, but also to cattle business, industry, commerce, hunting, and tourism." The proposal was received with skepticism by those who doubted the *ejidatarios* capacity of engaging successfully in such activities.[110] The 1971 Agrarian Reform Law gave form to this new policy allowing *ejido* participation in these activities. It also made it possible for two or more *ejidos* to form cartels or other types of mergers for this purpose.[111] FONAFE, the National Fund of *Ejido* Development, created years earlier by President López Mateos in 1959, was greatly expanded and given the role of promoter of these diversified economic activities.

In his relentless drive for public sector economic expansion, and being aware of the high profit potential of land when it becomes ripe for urbanization or for tourism, Echeverria decided to generate profits for the public sector by expropriating *ejido* land for these profitable purposes. The government justified these expropriations to the *ejidatarios* as a practical way of generating new employment opportunities for them in the tourism industry.

Such operations for profit purposes had been carried out in the past by politically influential private entrepreneurs through land barter operations with the *ejidos*, previous presidential approval, as required by article 146 of the older Code. Ironically, barter operations with private owners were outlawed in the 1971 Agrarian Law by the incoming Echeverria administration, after many *ejidos* had complained against FONAFE the government agency in charge of the barter operations (see case study 2 on the *ejido* Tlalpan). By eliminating the barter option the government was left without competitors in this field and FONAFE was put in charge of the program.

The government moved rapidly to promote *Fideicomisos Turisticos*, public-sector investment trust-funds to develop new tourist centers on *ejido* land on different attractive sites. Pressure was put on the *ejidatarios* on many of these sites to agree to the expropriation of their land.[112] Many of the projects administered by FONAFE ended in failure and accusation of fraud. When the next President José López Portillo (1976-82) ordered an audit of FONAFE in 1977, the results of that audit revealed widespread corruption. From its 452 enterprises, only 119—26% of the total—were profitable, while the rest were shown to lack any solid base. As a result, FONAFE was restructured under the name of FIFONAFE during the López Portillo administration.[113]

The state was not interested in promoting *ejido* independence and autonomy. "The Echeverristas had to maintain control of *campesino* mobilization in order to attain the goals of the national state" says Sanderson.[114] The practice of withholding individual parcel titles from most *ejidatarios*, keeping them in a state of permanent tenure insecurity in order to perpetuate *ejido* dependency is one clear example. The same attitude of treating the *ejido* has if it were state property and the ejidatarios mere tenants on that land, is of course obvious in the takeover of *ejido* land for commercial ventures undertaken by the public sector.

The Echeverria period is characterized by heightened conflicts for land in the urban and rural areas between the labor and the *ejido* sector. Affiliates of labor and bureaucratic organizations came into open confrontation with *ejido* interests over land in the urban peripheries and even clashed openly in the rural areas. Conflicts occurred between the CTM, the Workers Confederation and the CNC, the Peasants Confederation over land claims in rural areas.[115]

Although the tendencies of turning the *ejido* into an adjunct of the state took a much stronger form during the Echeverria period, no changes in the basic legal nature of the *ejido* property rights were introduced. The *ejido* remained the corporate property of the peasant communities and not government property. The new law did not move away from the parcelized *ejido* as the basic alternative open to the *ejidatarios* while insisting that "the individual parcels at no moment cease to be the property of the *ejido* nucleus of population."[116] The two types of *ejidos* which had evolved after 1917, the parcelized and the collective *ejido* , in spite of their differences, remained identical in this respect.

The 1984 *Ejido* Amendments

Agrarian legislation followed during the eighties the course that was set during the Echeverria administration, deepening the conflict of interests between the *ejido* and the state. Various amendments to the agrarian code were adopted in 1984 that dealt with the question of using *ejido* land for urban growth and the establishment of land reserves. These amendments, however, failed to remove the inequities standing in the way of using *ejido* land to implement the urban reserve and left largely untouched the major factors inducing illegal urbanization of the *ejidos* in the urban peripheries.

The question concerning the legitimacy of expropriating *ejido* land for urban development in general and for establishment of land reserves, and not only for public works, was settled in the affirmative. Articles 112-VI and 122-II of the Federal Agrarian Law explicitly included these objectives among the public interest causes which justify expropriation: "The following are public interest causes:.....I-VI. The foundation, improvement, conservation, and growth of the population centers whose ordering and regulation has been foreseen in the national, state and municipal plans of urban development and housing."[117] The urban development plans were declared to be the basis on which the expropriations would take place.

However, the terms of the expropriations were set in the same unfavorable terms that the *ejidatarios* had perceived in the past to be against their interests. The community could choose between relocation to another rural site in the country of similar agricultural quality and extension to continue their agricultural way of life or token compensation if they decided to stay in the city. If 2/3 of the community rejected relocation, which is the most common situation, they were entitled to receive as compensation a sum equivalent to the **agricultural** value of the land plus two urbanized plots per *ejidatario*. As to the profits of the land development operation the amendment assigned 80% of the profits to FIFONAFE while the remaing 20 % were designated for the expropriated *ejido*.[118] As already mentioned FIFONAFE would use these resources for economic development in other *ejidos*.

The Ministry of Urban Development and Ecology (SEDUE) or the DF government (DDF) were designated as the recipients of the expropriated lands.[119] Article 126 require that the objective of the expropriation be expressly mentioned in every expropriatory decree, explicitly authorizing these government agencies to subdivide, urbanize, and sell the expropriated land in order to prevent diversion of the land for any other purposes. If expropriated land was used for a different purpose than originally decreed, or more than five years had elapsed since expropriation took place without the land being used for its designated purpose, the land or its proceeds would not revert to the original *ejido* community, as Mexican legal doctrine would require for private property, but to FIFONAFE, the Fideicomiso Fondo Nacional de Fomento Ejidal. FIFONAFE, had a right to demand the reversion of the land in total or in part and would then be free to use the land for commercial purposes.[120]

In the past, *ejido* land had sometimes been expropriated for a specific public interest purpose, and then sold by the official institution at low prices to private interests for other purposes—most often commercial ones. Such irregularity had been severely criticized. Changing the use of expropriated land from its original purpose, regardless of whether these new purposes were legitimate or not, is illegal. It opens the door to serious suspicions of wrongdoing and mistrust of the process. However, opportunities for impropriety—far greater than before—were opened by article 126, proving again that *ejido* land is treated as if it were state-owned property. Previously, expropriation tended to be used as a cover to divert highly valuable *ejido* land to private interests. This constituted an illegal action. With the amendment of 1984, diversion of *ejido* land by state agencies could be accomplished completely within the framework of the law.

Stiffer controls and penalties were simultaneously introduced in the 1984 Agrarian Reform Law against individual *ejidatarios* alienating or even illegally leasing *ejido* land. Article 85-V established that any individual *ejido* member selling, renting, or in any way alienating his *ejido* plot, or tolerating that such actions be carried out with his *ejido* plot, would lose his membership in the *ejido* community and forfeit his rights to the plot. However, the penalties imposed on *ejido* authorities for similar offenses were much less harsh. Article 41-VIII established that any *ejido* official carrying out, promoting, authorizing, tolerating or failing to denounce the illegal transfer of *ejido* land, its leasing, or any form of illegal possession would have to be removed from his official post.

A further addition was made to the Agrarian Reform Law bringing to an end the long-held practice of using the ZUE as the framework for urban development: "It will be indispensable in every case to justify the effective need of creating or expanding the urbanization zone to satisfy preferentially the *ejidatarios'* own needs and not those of the nearby towns and cities."

Special amendments in 1984 also gave the *ejidos* some limited degree of freedom to associate with private enterprises for their mutual benefit. Addition "C" to article 138 allowed the *ejidos* to associate with private firms to harvest the *ejido* forests and process industrially the forest products. This freedom, however, was restricted to situations where the *ejido* was incapable of carrying out such a venture by itself or with government assistance, due to financial or technical limitations.

The *ejidos* were also authorized to establish long-term contracts with private timber processing industries to carry out continuous reforestation on their *ejidos*. The aim was to supply these industries with raw materials on a permanent basis. Mixed enterprises could be created for such a purpose with board members representing the *ejido,* the private firm, the Ministry of the Agrarian Reform SRA, the Ministry of Agriculture and Hydraulic Resources SARH, and the state government. Government bodies were also allowed to invest capital in these mixed enterprises.

Earlier permission to exploit *ejido* non-agricultural resources such as tourism, fishing and mining, was expanded.[121] Newly added article 144 allowed the *ejidos* to associate with third parties for these purposes, by way of limited-time contracts, subject to the approval of the *ejido* general assembly and the Ministry of the Agrarian Reform. The contracts, whose maximum period was not to exceed 3 years, could be renewed, taking into consideration the magnitude and recovery period of the capital investment as well as the record of the contracting party in complying with contract terms.[122] These amendments, allowing some limited form of cooperation with private enterprise, slowly opened up new ground in *ejido*/private partnerships, although they are still strongly rooted in the traditional approach of maintaining a highly dependent relationship between the *ejido* and the state.

Recent Developments in *Ejido* Legislation

Recent legislative developments, subsequent to this study, have taken place in Mexico which reflect an important reorientation in *ejido* policy. President Salinas de Gortari presented to the Chamber of Deputies on November 7, 1991 a proposal to amend article 27 of the Constitution and subordinate secondary legislation (Ley Agraria y Ley Organica de los Tribunales Agrarios) which was approved by a wide margin on February 20, 1992. The amendment has clearly moved in the direction of reducing the traditional relationship of dependency between the *ejido* and the state, granting the *ejido* a greater measure of autonomy and freedom.

The new legislation has finally settled the question on the fundamental nature of *ejido* property rights and laid to rest the attempts to reinterpret the *ejido* as state property, explicitly granting recognition to the *ejidos* and *comunidades* as distinct forms of private corporate property. In the words of President Salinas in his introduction to the proposal: "We confirm without ambiguity that the *ejido* and the *comunidad* are lands of the *ejidatarios* and *comuneros* protected by our Constitution."

The legal power of every *ejido* community to grant the individual *ejidatarios* their property titles over their individual parcels was recognized in consistency with the principle of *ejido* ownership and autonomy. The amendment also relaxed to some extent the restrictions on the marketability of the individual parcels and allowed the *ejidatarios* to transfer the property rights over their parcels to other members of the same *ejido* or community subject to restrictions that would prevent excessive accumulation or fragmentation of the land within the *ejidos*. The *ejidatarios* are now allowed to hire laborers to work their land. They can lease their parcels while working as sharecroppers or laborers, thus increasing their income.

The amendment also declared that the process of land distribution and establishment of new *ejidos* had come to an end as there was no more land available to be distributed: "We must recognize that the distribution of land that was mandated in article 27 of the 1917 Constitution has come to its final end." As a consequence the amendment revoked the constitutional clauses that required the government to grant land to petitioning peasant groups.

To attract badly needed capital to agricultural production and make possible more efficient scales of production as required by modern technology the amendment allows shareholding corporations to own agricultural land for their productive activities. Limitations on the maximum number of shareholders as well as on the size on the total extension of their landholdings were imposed on such corporations; these should not exceed as a group the proportional limits that apply to the small individual private owners. As anonymous shareholding was already outlawed a decade ago this was seen protection enough to prevent the formation of latifundia belonging to a single owner. *Ejidos* were also allowed to form shareholding corporations to attract investors contributing capital.

Specialized federal agrarian courts were mandated in the amendment to resolve the many legal questions related to limits and property rights claims by *ejidos* and communities. These judicial functions previously held by the executive branch were completely transfered to these courts. The amendment also repealed the constitutional clauses that created the Ministry of the Agrarian Reform as well as an host of other agrarian bureaucratic bodies including the *comisariados ejidales*.

Backing for the amendment came forward from most agrarian organizations. By a majority vote of 387 against 50 and two abstentions the Chamber of Deputies gave its approval to the amendment in 1992. Many prominent *agraristas* including the sons and daughter of Emiliano Zapata, Mateo, Diego and Ana publicly endorsed the amendment which was bringing the *ejido* nearer to the spirit of their father's ideology and away from the statist concept so alien to his libertarian spirit.[123]

* * *

Unambiguous conclusions can be drawn on the alleged claim that the *ejidos* are the property of the state given to the communities in usufruct. Although government often acted as if the *ejido* was state property, at no point in time, however, was there an actual legislative attempt to convert the *ejidos* into state property. Government was of course the power that carried out agrarian reform and distributed the land of the *haciendas* to the peasants. But in this operation it was clearly the catalyst but never the final intended recipient. The ideological discussion on the property rights of the *ejido* swung in the earlier beginnings between collective ownership by the peasant community and individual property by the peasants, but never around the possibility of state property. The 1992 amendment to article 27 is giving further recognition to the principle that the *ejidos* are the owners of their land and not the state. Much as the extreme advocates of a state-run economy would have liked this to be different, the legal definition of *ejido* as property belonging to the peasant communities and villages is too deeply embedded in the history and law of the country to be easily obliterated.

THE *EJIDOS* IN THE MAMC: AN HISTORICAL OVERVIEW

This chapter on the historic roots of the *ejido* institution would be incomplete without a reference to the origins of the *ejidos* and *comunidades* in the periphery of Mexico City. The origin of the Indian villages in the MAMC goes far back to the Aztec reign and the earliest period of the Spanish conquest. Although later depopulated through epidemics in the early colonial period, the valley of Mexico which surrounded the magnificent capital of the Aztecs, Tenochtitlán, was the most densely settled area in pre-Colombian times. A system of lakes in the lower lands of the valley encircled the city on whose outer shores were some of the most fertile lands producing during all the seasons of the year for the city markets. The Spanish crown granted many of the Indian villages in the valley collective property rights over some parts of this lands. The *comunal* lands that survived until this day in the region have their historical origin in the colonial grants to these villages.

Conflicts on land and water property rights between the village communities and the neighboring *haciendas* arose during the colonial period. These conflicts intensified during the independent period. Quoting Tutino on this point:[124]

> The *chinampa* cultivators in the communities just south of Mexico City not only retained their most valuable lands but maintained a strong community cohesion after independence, despite continuing attempts by city dwellers to gain access to their exceptionally fertile lands so close to the nation's largest market.

Some of the case studies in the literature refer to the conflicts that arose between the villages and the neighboring *haciendas* after the Lerdo Law began to be implemented in the second half of last century. During the Revolution many of the peasants of the southern DF villages joined the Zapatista movement whose center of operations was in the nearby state of Morelos. The claim that the owners of the *haciendas* in the central region were amongst the leading oponents of the Revolution can account for the fact that the area became an early target for the restitution and endowment programs implemented by the revolutionary governments.

Under the slogan "The promises of the Revolution are being fulfilled," Carranza carried out the first restitution in the country in 1916 even before the

Constitutional convention of Queretaro took place, returning 529.6 hectares of land to the village of Ixtapalapa. Other five restitutions in the DF area followed during the Carranza regime: San Juán Ixtayopan (August 30, 1917), Sta. María Cuautepec (March 9, 1918), Mixquic (April 14, 1918), Xochimilco (1918) and Sta. Ursula (1919).[125] Many of the lands of these communities are now occupied by irregular settlements and some of them are at present undergoing traumatic regularization programs.

The DF was also one of the earliest sites of *ejido* land grants in the country. Although the *haciendas* in the DF were among the smallest in Mexico, they were nevertheless among the first to be subdivided to carry out land grants to newly established *ejidos* groups. Most of the *haciendas* were typically between 200 and 500 hectares and only one which belonged to the Escandón family reached an extension of 2,563 hectares, an area still minor when compared to the great northern estates. The *haciendas* served as second homes to the landed aristocracy of the capital.

The overwhelming majority of the DF *ejidos* were created during the first historical period of the agrarian reform, soon after the Revolution, and not during the Cárdenas period as erroneously assumed by some authors. The largest agrarian reform program in the DF to establish *ejidos* was carried out during the presidency of Alvaro Obregón (1920-24). He distributed 141 sq. km., almost half (48%) of the total area of *ejido* land in the DF, to 10,321 *ejidatarios* (table 15).

The *haciendas* in the area suffered great damages during the revolutionary struggle. Their livestock and equipment were lost and their land lay fallow as their laborers were drafted into the federal army or fled to the many revolutionary armies. Some *hacienda* owners in the DF managed to divide early enough their estates into small properties within the limits allowed by the agrarian legislation in order to forestall land distribution. Many of the small private properties, *pequeñas propiedades*, to be found in the region trace their origin to these subdivisions or to the minor fractional parts of the *haciendas* that were left with their original owners when these were expropriated. At the end of that first period (1934) as much as 216 sq. km., 74% of the total land grants in the DF had already been distributed. A total of 66 *ejidos* were created in that period for 19,835 *ejidatarios*.

Many of the existing *ejidos* which were generally small petitioned the government for more land in the years of President Lázaro Cárdenas (1934-40). These petitions were sometimes granted by additional expropriations from the same *haciendas* of which the original grants were carved out. President

Cárdenas expanded 23 of the earlier *ejidos* and created 8 new *ejidos* in the DF. He distributed a total of 68 sq.km., 23.4% of the total, to 2,194 *ejidatarios* (table 15). Thus by 1939, at the end of his administration, 97.3% of the area of the *ejidos* in the DF had already been distributed. With one minor exception, *ejido* formation in the DF practically ceased at the end of the Cárdenas administration.

Table 15. *Ejido* Land Grants (sq. km.) by Presidential Periods in the DF (1915-1961)

Presidential Period		*Ejido* Land Grants			
		Area	%	Ejidatarios	%
First Period:					
Carranza	(1915-21)	25.7	8.8	4,949	21.4
Obregón	(1921-24)	141.0	48.5	10,321	44.7
Elias Calles	(1925-28)	8.1	2.7	1,189	5.1
Pórtes Gil	(1929-30)	39.6	13.6	3,349	14.5
Ortiz Rubio	(1931-32)	1.1	0.4	27	0.1
A. Rodriguez	(1933-34)	0.0	0.0	0	0.0
Subtotal	(1915-34)	215.8	74.0	19,835	85.8
Second Period:					
L. Cárdenas	(1935-40)	68.2	23.4	2,194	9.5
Third Period:					
Avila Camacho	(1941-46)	0.0	0.0	0	0.0
Alemán V.	(1947-52)	6.0	2.1	888	3.8
Ruíz Cortines	(1953-57)	0.0	0.0	0	0.0
López Mateos	(1958-61)	0.6	0.2	140	0.6
Subtotal	(1941-61)	6.6	2.3	1028	4.4
Total	(1915-61)	290.4	100.0	23,057	100.0

Source: Cruz Rodriguez Ma. S. (1981)

Taken as a whole, the total agrarian reform program in the DF since 1916, created 82 new *ejidos* for 17,261 *ejidatarios* and carried out restitutions to 11 *comunidades* which held 8,567 *comuneros* (table 16). *Ejidos* and *comunidades* together covered a total of 741 sq. km., almost half of the total area of the DF (table 15). Two thirds of the total land distributed (500 sq. km.) went to the *comunidades* as restitution while one third (242 sq. km.) went to the *ejidos*. As a result, the acreage per community (4526 has) and per peasant (5.8 ha.) was much higher for the *comunidades* than for the *ejidos*: 4526 has/*comunidad* and 5.8 ha./*comunero* against 296 ha./*ejido* and 1.4 ha./*ejidatario* (table 16), a situation which is also characteristic for the State of Mexico and the country as a whole.

Table 16. Comunal Land Restitutions and *Ejido* Land Grants in Mexico, the DF, and the State of Mexico (1915-1981)

	Comunidades	*Ejidos*	Total
Total Country			
No. of Units	1,486	23,589	25,075
Area (Sq. Km.)	157,046	826,360	983,406
Hectares/Unit	10,568	3,503	3,922
No. Members	354,697	2,908,963	3,263,660
No. Members/Unit	365	123	130
Hectares/Member	44	28	30
DF			
No. of Units	11	82	93
Area (Sq. Km.)	498	243	741
Hectares/Unit	4526	296	796
No. Members	8,576	17,261	25,837
No. Members/Unit	780	210	278
Hectares/Member	5.8	1.4	2.8
State of México			
No. of Units	99	1,038	1,137
Area (Sq. Km.)	2446	8,621	11,067
Hectares/*Ejido*	2470	830	973
No. Members	19,423	168,445	187,868
No. Members/*Ejido*	196	162	165
Hectares/Member	12.6	5.1	5.9

Source: Zaragoza J.L. & Macías R. (1980), CENIA (1981)

The average size of the peasant communities created by restitution or by new grants were the smallest in the whole country, 796 ha. per unit as against 3,922 ha. in the nation, while the number of peasants assigned to each unit was among the highest, 278 members against a national average of 130 (table 16). This determined a much lower average acreage per *ejidatario* in the DF than in the country, 2.8 ha. against a national average of 30.8 ha. The size of the individual parcels was only a small fraction of these areas, due to the common areas. The endowments in the State of Mexico were higher than those of the DF but still far below the national average. As a result of these ratios the economy of the peasants in the metropolitan periphery is characterized as what is described as *minifundismo*, token size agriculture which can hardly provide a living for its members given the minimal capital resources with which they operate.

When the agrarian reform came to an end in the DF and *ejido* formation ceased, a new era of *ejido* expropriations for urban development purposes was just about to begin as a result of rapid urbanization. President Cárdenas abstained from expropriating or using *ejido* land for urban purposes. His immediate successor, President Manuél Avila Camacho (1940-46), was the first to privatize *ejido* land for urban development purposes in the MAMC. He expropriated *ejido* land for industrial development to create the Vallejo industrial zone. As industrialization and urban growth in the MAMC became top priorities in government policy the conversion process of *ejido* land for urban purposes became a most important question. Barter of *ejido* land for private agricultural land which could provide for the relocation of the *ejido* community was made legal in the 1942 Agrarian Code, subject to a 2/3 acceptance vote by the community and presidential approval. The land barter clause allowed the development of a number of upper class private residential areas in the MAMC, Ciudad Satelite and Jardines del Pedregal among others, until it was repealed by President Echeverria in 1970.

Like the barter clause which was instrumental in opening up the suburban *ejidos* for private development purposes, the *zona urbano ejidal* clause introduced in 1954 by President Ruíz Cortines played a similar role in opening up the *ejidos* to low-income urbanization.[126] The decree which allowed the sale of excess parcels in the *Zona Urbano Ejidal* to non-*ejido* members wishing "to settle within the *ejido* villages to cooperate with the effort to develop them," led to the creation of the *ejido* settlements in which hundreths of thousands of irregular settlers found refuge. The extension of *ejido* and *comunal* land in the urban periphery began to diminish at an increasing pace through these means and through presidential expropriation for public interest purposes.

According to the recent data of the 1991 Agricultural Census (VII Censo Agropecuario) and the 1988 *Ejido* Survey (Encuesta Nacional Agropecuaria Ejidal) the *ejidos* and *comunidades* in the DF still hold a very significant area of 662.1 sq. km., 44% of the total area of the DF. Over 70% of the land belongs to the *comuneros*; 42.4% is classified as agricultural land, 38.7% as forests and 11.8% as grazing land. There are now 38 peasant comunities left in the DF with a total membership of 20,373 *ejidatarios* and *comuneros*. Of them, 14,345 are parcel holders whose individual parcels measure an average of 1.38 hectares. Nineteen *ejidos* and *comunidades* are the custodians of the forest lands and two of them draw their living from harvesting their pine and oyamel trees. After 76 years of agrarian reform the tenure situation is still uncertain in many of the communities and the *ejidatarios* continue struggling in them for the confirmation of their property rights.

Although the *ejidos* hold such highly valuable resources of land and forests in the most dynamic area of the nation, the data reveal a dismal picture of poverty and backward agricultural technology. There is practically no irrigation on the totality of the land of the peasant communities in the DF, seemingly due to the higher priority of urban consumption over agricultural production and the high cost of water in the MAMC. In almost every indicator these peasant communities stand below the average of the *ejido* comunities in the nation. Only 6 of the 38 *ejidos* receive credit from the official agricultural bank Banrural, the FIRA programs or the private bank system while the rest must turn to private high-interest lenders. Most of them do not use improved seeds (35), lack tractors (27), nor benefit from any technical assistance (26). Many of them do not use pesticides (18) nor fertilizers (13). One can hardly escape the conclusion that a low priority is given in public policy on providing stronger backing to the processes of agricultural production in the rural periphery of the MAMC.

It can be safely argued that such a situation in the rural periphery provides the *ejidatarios* with strong incentives to sell their land even when that is against the law. Although it is true that even highly developed farming can rarely compete with the rising land values in a situation of urban expansion, minimal productivity certainly weakens any possible defenses against such behavior. In parallel with their technological backwardness as agricultural producers the *ejidos* in the DF lack urban infrastructure and face a situation of a low level of services. Twenty one of the *ejidos* lack potable water, 18 of them lack electricity and only 3 of them have pavimented access roads. Such a situation contributes to their eventual urbanization as irregular low income settlements. While isolation, lack of services and legal non-marketability act as a deterrent to the entrance of the higher income groups of people into these *ejidos* it depresses the sales value of the land facilitating the entrance of low income settlers who are not deterred by these factors.

---

1 (Castells, 1981:p. 138).

2 (Schteingart 1981, p. 169, 170; 1983:p. 281).

3 (Cornelius & Kemper, 1978:p. 72; Garza & Schteingart, 1978:p. 72).

4 (García Peralta, 1983:p. 1439).

5 (Ward, 1983:p. 35; 1981:p. 61).

6 (Stavenhagen, 1986).

7 (SMP Sociedad Mexicana de Planificación. Tlaxcala Conference Acts, 1982:p. 37-51)

8 *Recopilación de las Leyes de los Reynos de las Indias*. Ley 1, Titulo XII, Libro IV. Madrid, 1943.

9 (Senior, 1958:p. 24).

10 (Chevalier, Francois; Sanderson S. 1981; Simpson, 1952; p. 20).

11 The legal scholar Wistano Luís Orozco objected soon after in "Los Ejidos de los Pueblos," to the imprecise use of the term.

12 (Whetten, 1948).

13 (Molina Enríquez, 1937).

14 (Tutino, 1986:p. 245; 257).

15 (Ibid: p. 258).

16 (Ibid:p. 261, 262, 274).

17 (Ibid:p. 263).

18 (Ibid:p. 264-265).

19 Historian Jean Mayer counted around seventy such rural rebellions during the XIX century.

20 (Díaz Ramírez, 1938).

21 (Tutino, 1986:p. 264-265).

22 According to the 1910 census, 840 *hacendados* owned 50% of the land, while the rest was owned by 400,000 *rancheros* working family-size farms.

23 (Whetten, 1948). Twenty percent of the entire area of the country passed into the hands of land surveying companies. They received one-third of all vacant lands they located and surveyed. Many of the vacant lands belonged to village communities, on which lay the burden of proof of their ancient property rights. These companies ended up with enormous estates of millions of hectares in size in the sparsely settled areas of the north of Mexico. By 1910 approximately 27% of the total area of the country was transferred from the public domain to a few private hands, for a total income of $12 million MP.

24 "The telegraph network was extended to remote areas, increasing its size sevenfold. Coal was being extracted and there was an oil glut in Tampico on which the British planned to run their navy. Shiploads of steel and machinery docked at Veracruz, going to the mining camps, railroads, oil fields and the textile factories being built in the center of

the country" (Hodges, 1983).

25 To the traditional exports of gold, silver, and copper were added lead, zinc, and other industrial minerals.

26 Sugar in Morelos, hemp in Yucatán, cotton, chickpeas and vegetables in Sonora and Coahuila, tobacco, coffee, and vanilla in other states.

27 (Tutino, 1986:p. 269; Whetten, 1948).

28 (Whetten, 1948).

29 "In the rich sugar-producing area of the State of Morelos, the *hacendados* bent on increasing their acreage, steadily dispossessed the peasants of their land between 1880 and 1910. Whole villages disappeared, as residents were driven of their communal lands to work as day laborers on the *haciendas*. The *haciendas* which suffered from a chronic labor shortage problem, used permanent indebtedness to the *hacienda* store to keep the now landless peasants tied to the *haciendas*, thus virtually reintroducing serfdom in the countryside" (Niemeyer, 1974:p. 8).

30 (Tutino 1986:p. 275 q. Stevens, Agrarian Policy : p. 162-163). Jean Meyer indicates that between 1876, when Díaz came to power and 1901 there was at least one agrarian uprising each year. From 1902 to 1910 only one occurred in 1905.

31 A combinations of factors contributed to the downfall of Díaz: The great foreign debt that the nationalization of the railroads in 1903 brought about; the extreme shortage of credit created by the shift from the monetary silver standard to the gold standard in 1905; the interrupted infusions of foreign capital due to the Wall Street financial crisis of 1907, all played (Tutino, 1986: p. 333-335).

32 While commercial agricultural exports expanded quickly at an average annual rate of 2.5%, agricultural production for the national market grew at 0.65 %, and food crop production actually declined at an average annual rate of 0.5%, for approximately 20 years.

33 Years earlier the Mexican Liberal Party which stood under the influence of socialism and the anarchistic ideas of Ricardo Flores Magon, issued in St. Louis, Missouri a political program in which they expressed the need for land reform, protection of the Indian, and revision of the Constitution. Basic MLP proposals were later written into the revolutionary Constitution .

34 Cabrera a law-firm associate and friend of Molina Enríquez was deeply influenced by him in the agrarian question.

35 "The creation and protection of agrarian property is a problem of great importance to protect the smallholder against the great property-holders. To accomplish this, it is urgent to undertake throughout the country a series of reforms leading toward equal footing for great and small rural property." (Cabrera, Address before the Chamber of Deputies, 1912).

36 (Senior, 1958:p. 25).

37 Ibid.

38 (Katz, 1981).

39 (Womack, 1974; Guerrit Huizer, 1968).

40 "Free land, free parcel, free man. Without overseers and bosses within the ejido, without individual tyrannies but also without tyrannies from the state and the collectivity. " As quoted from A. Soto y Gama by historian E. Krauze (La Jornada 09/24/91).

[41] (Senior, 1958:p. 24-25).

[42] Codificacion de los decretos del C.Venustiano Carranza. Mexico D.F. Imprenta de la Secretaria de Gobernacion, 1915. (q. Niemeyer, 1974.)

[43] (q. Sanderson, 1981).

[44] Villa also issued his own land reform decree on May 1915. The lands were to be placed initially under control of the state governments to obtain revenue to finance the revolutionary struggle. All estates above a certain maximum size were to be divided among the peasants after the revolution, to create small private landholdings with full private property rights, that would include the right to sell, transfer, or mortgage. Compensation to the *hacienda* owners would be paid by the peasants themselves by means of a system of small installment payments. Before distributing the land, the state governments would be authorized to use some, to restore to the villages the lands they had lost, to secure unpaid taxes by the *haciendas*, and to finance pensions for widows and orphans of revolutionary soldiers. (F. Katz, 1981).

[45] The Agrarian Commission later became the *Departamento Agrario*, then the *Departamento Agrario y de Colonización* (DAAC) in 1958, and was finally raised to ministerial rank in the early seventies under the name *Secretaría de la Reforma Agraria.*

[46] Cincuentenario de las adiciones y reformas al Plan de Guadalupe, del 12 de diciembre de 1914. (q. Niemeyer, 1974.)

[47] The maximum size was left undefined.

[48] Pastor Rouaix, Carranza's link to the convention expressed his regret years later for having chosen this approach, which he wished could have been formulated as emanating directly from the people to the state (Rouaix, 1984:p.143).

[49] Ibid. See A. Molina Enriquez (1911, re. 1978-91:p. 474).

[50] Ibid.

[51] (Art. 27-VI).

[52] Delegate Hilario Medina, objected to the wholesale nullification of all the transactions, including those completely legal which were made with the full consent of the parties involved. Mugica responded that although many transfers of property were legally made, the result had been disastrous for the rural communities: the problem was not how something had been done but what its consequences had been. Diario , p. 1117.

[53] (Art. 27. Constitucion Politica de los Estados Unidos Mexicanos de 1917).

[54] (Art. 27, VII-3).

[55] (Art. 27, VII-9). The *ejido* concept which is rooted in this clause was later defined in the Agrarian Code to mean "the total area of land endowed to a population nucleus, including the agricultural lands, the urbanization zone (the later ZUE), the school plot, and any other type of land different from its agricultural land, necessary to satisfy the collective needs of the population nucleus" (Código Agrario, Art. 80, 1969).

[56] (Niemeyer, 1974:p. 145).

[57] Diario de los Debates, II, 1225.

[58] Diario de los Debates, II, 1188.

[59] (Art.27, p. VI. Constitucion Politica de los Estados Unidos Mexicanos de 1917).

[60] (Tutino, 1986:p. 261-262; 347).

[61] The legislation which was produced in the first period includes the Law of *Ejidos* of 1920 (Ley de Ejidos), the agrarian bylaws of April 10 of 1922 (Reglamento Agrario), the

Law of Parcelized *Ejido* Property of 1925 (Ley de Patrimonio Parcelario Ejidal), and the Law of Endowment and Restitution of Lands and Waters of 1927 (Ley de Dotaciones y Restituciones de Tierras y Aguas). The second period produced the first two Agrarian Codes of 1934 and 1940. The third period produced the Code of 1942 and its additions of 1954. The fourth period produced the Law of the Agrarian Reform (Ley de la Reforma Agraria) adopted in 1971 during the Echeverria administration, and its various subsequent additions, including those of January 1984. A new code is being adopted in 1992.

62 The maximum size allowed in the law for private land property in this period was 150 hectares for irrigated land.

63 In the early years of this period, the Comisión Nacional Agraria recommended a collective form of agricultural production, without parcelization, in a circular (Circular 51) dated October 11, 1922, on grounds of technical efficiency and social interest. Little is actually known on the implementation of this initiative and its results which remained on paper (Aguirre Avellaneda, 1976).

64 (DO, Dic. 30, 1925: Ley de Patrimonio Parcelario Ejidal).

65 (Circular 22, Comisión Nacional Agraria, 1917; q. Avellaneda, 1976).

66 (Simpson, 1937; 1952).

67 (DO, April 9, 1926).

68 (Aguirre Avellaneda, 1976; Reforma Agraria en México, DGE. México, 1937).

69 This national average, which also includes the larger northern *ejidos*, obviously does not reflect the actual much smaller size of the individual agricultural parcels, as it includes the non-agricultural grazing lands and the other collective lands of the *ejidos*.

70 (Durán, 1967:p. 119).

71 (Aguirre Avellaneda, 1976:p. 122).

72 DO, April 12, 1934. Reforma del Código Agrario de los Estados Unidos Mexicanos, April 9, 1934.

73 DO, August 12, 1937. (Sanderson, 1981:p. 107).

74 DO, October 6, 1936; August 9, 1937.

75 (Aguirre Avellaneda, 1976:p. 58).

76 Codigo Agrario de los Estados Unidos Mexicanos, September 23, 1940.

77 Ibid.

78 (Durán, 1967:p. 766).

79 DO, December 20, 1935.

80 (Simpson, 1937).

81 (Tutino, 1986:p. 347).

82 (Grindle, 1986:p. 176).

83 (Aguilar Camín & Meyer, 1990:p. 172).

84 85% of all ejido plots were subsistence oriented, while only around 50% of the private farms was in that category in 1960 (Hansen, 1971:p. 62).

85 Codigo Agrario de los Estados Unidos Mexicanos. Diario Oficial, April 27, 1943.

86 (Aguirre Avellaneda, 1976:p. 61; Sanderson, 1981:p. 144).

87 (Hansen, 1971:p. 60-61).

88 (Aguirre Avellaneda, 1976:p. 59-62; Sanderson, 1981:p. 144-145).

[89] Paradoxically, the advocates of collective ejido farming saw the root of the problems of the small *ejido* parcels in their refusal to accept collective organization (Aguirre Avellaneda, 1987).

[90] (Aguirre Avellaneda, 1976:p. 59-62; Sanderson, 1981:p. 144-145).

[91] (Eckstein Salomón, 1966).

[92] It is argued (Aguirre Avellaneda, 1976) that *ejido* parcels only reached an average of 5.5 hectares at the end of the third period, not being able to break away from the basic pattern of the Cabrera *minifundio.* Only 18% of the land received was plowable land, the rest being grazing land, forests, and unproductive soil kept as collective property.

[93] V Censo Agricola, Ganadero y Ejidal, 1970. By 1960 the ejidos held 40% of the total of irrigated land in the country, IV Censo Agricola Ganadero, 1960.

[94] The administrative process from the initial request for land to the granting of individual tenure title, where it ever occurs, takes 13 years on average according to an official survey of the Ministry of Agrarian Reform SRA. Stavenhagen refers to *ejidos* where the process is not completed after 20 years, and the second generation of *ejidatarios* has to carry on the struggle for individual title that their deceased fathers began.

[95] (Stavenhagen, 1986).

[96] Quoting Hansen (1971): "Employees of the Banco Ejidal are more often accused of fraud and corruption than those of any other Mexican governmental agency. And in Mexico, being number one in corruption is a very high fence to ride at. Seldom does a year go by without the uncovering of a major scandal in the bank; charges of fraud at the local level are a weekly occurrence."

[97] (Puente Leyva, 1975).

[98] (Sanderson, 1981)

[99] (Newell & Rubio, 1984).

[100] Ibid.

[101] (Gentelman, 1984).

[102] (Aguirre Avellaneda, 1976).

[103] The high hopes placed on the agricultural and agrarian reform policies of this period were not realized. The annual agricultural production growth rate fell to 1.6%, below that of the 1965-70 period (2.3%), drifting further away from the demographic growth rate (3.6%). From food self-sufficiency before 1970, agricultural imports of basic foods climbed to 22.5% of Mexico's balance of trade in 1973. A deterioration of the trade and foreign exchange balance took place. The share of GDP of the agricultural sector still supporting 40% of the country's population, dropped to less than 10% (Gentelman, 1984). Even more critical were the effects on the general economy. The end of the long era of cautious frugal lending of previous decades undermined the spending capacity of the state and initiated an inflationary spiral in 1972 which raged at 60% by 1976; a fall of exports due to inflation and a deterioration of the trade and currency balance; a sharp decline in private investment reached critical proportions by 1976.

[104] (Sanderson, 1981:p. 159-185).

[105] (DO, March 16, 1971).

[106] That area is around half of the land distributed by his predecessor, Díaz Ordaz. With López Portillo (1976-82), who distributed 56,089 square kilometers to 210,594 peasants,

land distribution returned to the slower pace that was characteristic of the the first 18 years of the previous period. By 1981 there were in the country 23,589 *ejidos* and 1,486 communities holding 983,406 square kilometers of land (tables 9 and 10).

107 While the government claimed credit for the creation of thousands of them, it actually only issued a decree creating 634 collective *ejidos* one day before the President left office (DO, November 30,1976; Warman A.; q. in Sanderson, 1976).

108 (D.O. April 5, 1976).

109 (Sanderson, 1976).

110 Cosio Villegas, Daniel. "El Estilo Personal de Gobernar".

111 The new credit law granted the *ejidos* and communities corporative personality as recipients of bank credits, and not only to their credit societies as before. The same recognition was extended to *Uniones Ejidales*, cartels grouping various *ejidos.*

112 Quoting Sanderson: "The Union Ejidal Bahia de Banderas made up of seven *ejidos* (where one of the major tourist projects was instituted), told of the extortion and pressure by the state and federal governments to give up their lands to tourist-resort developers and government functionaries." (Sanderson, 1981:p. 184).

113 (Lugo Medina & Bejarano, 1981).

114 (Ibid. p. 182).

115 (Sanderson, 1981).

116 (Ley Federal de la Reforma Agraria, Art. 58. D. O. March 16, 1971).

117 (DO, January 17, 1984; Ley Federal de la Reforma Agraria, 1984).

118 (LFRA art. 122, I-II).

119 (LFRA, art. 117).

120 (LFRA, art. 126).

121 (article 185).

122 The modern-day involvement of the *ejidos* into non-agricultural activities such as tourism, industry, fishing, and mining was recently challenged as unconstitutional by a leader of the opposition party, Accion Nacional, Manuél Gonzalez Hinojosa. He argued that *ejido* land was expropriated from the *haciendas* for agriculture, cattle raising, and forestry. To divert it for non-agricultural purposes invalidates the original expropriatory act, justifying its return to its original owners–the *haciendas.* (González Hinojosa, "Ejidos Turisticos, Absurdo Antijuridico"; Excelsior, August 26, 1989).

123 Ana and Diego Zapata declared: "Zapata would have backed the present amendments to article 27." (Excelsior,April 11, 1992).

124 (Tutino, 1986:p. 235; Lira, "Comunidades Indigenas" pp. 21, 77-78, 117-119).

125 (Crúz R., 1981).

126 (Diario Oficial, March 25, 1954).

Chapter V

THREE CASE STUDIES

The three case studies on irregular settlements in the MAMC summarized in this chapter were chosen from the literature to illustrate in a more tangible and specific manner the questions that the MAMC confronts in its periphery. They reflect the role played by the highly restricted property rights of the *ejidos* on the character of development in the MAMC periphery, providing convincing direct and indirect evidence for the thesis suggested in the previous chapter.

These case studies are not limited to one specific area, but span the two political entities which integrate the MAMC, the Distrito Federal and the State of Mexico, with their different political environments. The San Bernabé and Miguél Hidalgo case studies are situated on the southern periphery of the Distrito Federal being typical of the irregular settlements in Tlalpán, Ajúsco, Contreras, Coyoacán, Xochimilco, Ixtapalápa and the other *delegaciones* to the south of the DF, where many hundreds of thousands of settlers established themselves on *ejido* land under the ZUE cover. The Netzahualcoyótl study, on the other hand, is located in the State of Mexico to the northeastern side of the MAMC, where the greatest concentration of irregular settlers in company sponsored developments partly on *comunal* land took place on the desiccated bed of Lake Texcoco. Together they represent a combined population of at least five million people.

The San Bernabé case study belongs to Jorge Durand[1] and was brought up to date by including later legal developments published in the Diario Oficial (1989); the Miguél Hidalgo study is the original work of María Luisa Medina Lúgo and Fernando Bejarano González (1981);[2] the Ciudad Netzahualcóyotl study draws from various authors such as María Teresa Guerrero et al., Martín de la Rosa, and Martha Schteingart.[3]

The introductory part of each case study refers to the early historical origins of the property rights of the land on which the irregular settlements are situated. The early property rights problems are often the basis of the land conflicts that arose in later days. The *haciendas*, the large land holding estates, and the free Indian villages, the owners of *comunal* land, were already in colonial days involved in property rights conflicts which lingered on for centuries. These conflicts brought about the early implementation of land reform in the rural periphery of the capital soon after the Mexican Revolution, as explained in chapter IV.

The case studies document the actual implementation of the land reform program in the area of the capital after the Revolution. Partition of the neighboring *haciendas* resulted in the creation of new *ejidos* or in the restitution of *comunal* land to the old village communities in the vicinity of the city, mainly during the twenties. Future conflicts were incubated in this period of token agrarian reform. In the Netzahualcoyótl case, restitution of the area to the local *comuneros*, as required by the Constitution, was held back and used instead to carry out private land endowments to influential veterans of the revolutionary struggle.

The city's rapid territorial expansion after that initial period as a result of its demographic and economic growth, provides a more recent historical reference point. The expanding urban area advanced rapidly to the suburban *ejidos*, whose existence as agricultural rural communities was coming to a rapid end as the city reached their doors. The effects that the city's rapid expansion produced on them, particularly since 1940, are portrayed: a change in the occupations of many *ejidatarios* from self-employed land tillers within the *ejidos* to external day-laborers or petty merchants in the city, a decline in the productivity of their agricultural activities, and a rapid increase in the commercial value of the land.

The legal system in which the *ejidos* are enmeshed constitutes the background of the case studies. As these *ejidos* disintegrate as units of agricultural production they confront an inflexible legal system that offers dismal alternatives for their incorporation to the city. The inefficiencies and inequities of the legal means available to achieve *ejido* conversion lead them to illegal escape routes to evade the otherwise intractable *ejido* restrictions, such as sale under the ZUE cover, settler invasion, and fraudulent development.

The irregular settlement process itself is obviously the main focus of the case studies. The case studies open a window from which to experience the actual process of irregular settlement formation in the *ejidos* and its high social costs. Government policy and its agents can be observed in full operation from its early implementation: the local authorities, the settler organizations, the party, the Oficina de Colonias. The endless struggle for property titles and services that spans decades, and the role of *ejido* illegality as an effective excuse for postponing the service demands of settlers for many years. The dependency and helpesness of the settlers in this respect and the enormous amount of social energy spent in barren conflicts. The evidence of the settlers' desire for legal title to their plots is unmistakable, hardly leaving room for any arguments about their intense wish to escape illegality and undefined property rights, disproving those authors which claim otherwise.

The regularization programs are the culminating finale of the case studies. The regularization programs of CORETT described in the case studies leave little doubt that the agrarian bureaucracy saw the suburban *ejidos* as an easily available resource of great value obtainable by symbolic compensation. The other side of the coin, the reaction of the *ejidatario* to government expropriation in the context of the declining agricultural value and rising commercial value of their land is no-less well reflected. Their fierce opposition to the terms of the regularization programs and the rationale of that opposition become clear. The objections of the *ejido* communities to the expropriation process and their powerlessness vis-a-vis government is apparent in the unending conflicts described.

In conclusion, the case studies substantiate the argument that the *ejidatarios* are not to be perceived as the culprits but rather the victims of an inequitable and inefficient system of *ejido* land conversion with no benefits for them and high costs to society.

CASE STUDY 1: THE *EJIDO* OF SAN BERNABÉ

The San Bernabé case study by Jorge Durand surveys the land question of an *ejido* located at the southern periphery of the DF, from its early historical origins to the present. It offers valuable insights into the process of irregular settlements formation on its land after the Uruchurtu ban in 1953 under the cover of the ZUE fiction. The case study provides a most valuable account of the regularization program in the *ejido* and of the reactions of *ejidatario* and settlers, which should serve as valuable evidence for the explanations developed in chapter III.

The *ejido* of San Bernabé is located in the *delegación* of Contreras within the DF, to the south of the city at a short distance from the UNAM campus, the upper-class development Jardines del Pedregal, and the two beautiful national parks, Desierto de los Leones and Dinamos de Contreras. It stands on an elevated area with nice views and minimal air pollution, in the vicinity of a ravine through which currents of fresh air make their entrance.

Historical Origins of the *Ejido* San Bernabé

San Bernabé Ocotepec is the ancient Pre-Columbian Indian village of the *ejido*.[4] The nearby *hacienda* La Cañada which was originally established around 1750, expanded on communal land belonging to the village. Dispossession of the villagers from their land served the hacienda's interests by making available the labor force needed for its operation.[5] Gradually the *hacienda* grew to 1100 ha., surrounding the village on all sides. The village possessions before the revolution were reduced to 29 hectares of agricultural land, and 373 hectares of communal woodlands with a rich variety of valuable hard and softwood trees on a hill historically known as Cerro del Judío.

The village population, whose livelihood depended on their direct labor on the *hacienda*, consisted of around 130 families during harvest time.[6] As they could not draw a living from their 29 hectares of land, they leased another 116 hectares from the *hacienda* and used part of their crop and labor to pay for the lease. Another source of income of the village was the communal forest. Some of the villagers worked in textile and paper mills that existed in the nearby towns of Contreras and Tizapán.[7]

The *hacienda* seems to have ceased cultivation during the Revolution (1910-21). In 1923, when it was sold, it was abandoned and in ruinous condition. The field laborers had taken away the cattle and had taken possession of the orchards. The village itself suffered during the revolution from military operations. Many village huts were burned down, and the village youth was drafted into the federal army.

After the Revolution, on November 16, 1920, the village presented a petition for land restitution to the Federal Government, on grounds that: "President Porfirio Díaz (1876-1910), dispossessed us of our land by force." As proof of their claims, they presented 4 notebooks which contained maps and property titles of their *comunal* lands. They made reference to an ancient cloth, now under custody of the local church, in which the original boundaries of the *comunal* land belonging to the village community were drawn. A few months later, (March 29, 1921) they tried to withdraw their petition for undisclosed reasons. The agrarian authorities interpreted the withdrawal as the result of pressure from the landowners, and refused the withdrawal request. On October 24 of that same year, they renewed their original claim of land restitution.

The village documents were examined for authentication. An expert paleographer dismissed the documents as apocryphal on grounds that neither the ink nor style corresponded to the period claimed. As a result of this setback, the village presented a new petition on August 23, 1922, in which they stated: "In view of the paleographers assertion, we are unable to prove when our lands became private property, which we assume must have occurred before 1856." Therefore, instead of a restitution request they opted for simply petitioning the government for an endowment of land as a new *ejido*, in accordance with the legislation adopted by the revolutionary government.

The *hacienda* was sold in the meantime by its owner (November 26, 1923) to three partners for $MP 45,000 , under the understanding that none of the claims against the *hacienda* included the *hacienda* house and the higher quality land. On March 25, 1924, three years after they initiated their claim, the village of San Bernabé was finally granted 383 hectares of agricultural and forest land from the *hacienda*.. The land was divided among 143 petitioners[8] who received on average, parcels of 3 hectares. Only half of the land received was cultivable land, while 123 hectares were hilly forest, which the *ejidatarios* were required to maintain as communal woodlands. The *ejido's* economy was basically one of subsistence agriculture, supplemented by income from a few marketable products such as pulque, flowers, fire wood, and coal from the communal forest. As a whole, their income from agriculture overshadowed any other source.

Two nearby villages, La Magdalena and San Bartolo, also petitioned for land from the same *hacienda* La Cañada. La Magdalena received a land grant of 135 hectares and San Bartolo 77 hectares. The *hacienda* was left with 418 hectares, in two separate locations. Two years later (December 27, 1926), the owners of the *hacienda* received $25,260 MP as compensation for the lands taken for San Bernabé, in addition to the compensation for the land taken for two other villages.

Ten years after the establishment of the San Bernabé *ejido* (1934), when President Cárdenas came to power, the *ejidatarios* of San Bernabé and San Bartolo petitioned again for the expansion of their *ejidos* by further partition of the *hacienda*.. A new agrarian census was undertaken on April 6, 1937. This count found that the names of 56 *ejido* members from the 1924 land grant were missing. At the same time, 168 new names made their appearance in the census of land petitioners.[9] The petition for a new land grant was rejected on grounds that it would be against the law to further partition the *hacienda*, considering that its size was within the legal limits, and had already gone through previous partitions.

The rejected petitioners were, however, recognized as a newly constituted agricultural population nucleus, which meant that they were entitled to form a new agricultural settlement anywhere where land became available.

Ninety six *ejidatarios* received official documentary recognition in 1941 of their tenure rights in the *ejido*. Another group was left waiting for recognition of their rights, creating a split in the community which lasted for many years with serious consequences.

Soon the *hacienda* found itself threatened from another side. Local textile workers began to pressure the industrialists in Tizapan for housing facilities, as required by Article 123 of the 1917 Constitution. Although these industries maintained housing barracks for their workers, only 35% of their workers had access to them. In a manner typical of the period, the workers of the textile mills began to prepare themselves in 1940 to invade the land of the *hacienda*. To protect his land, the owner of La Cañada lent his land to the *ejidatarios* of San Bernabé and San Bartolo for cultivation, on the assumption that the *ejidatarios* would hardly allow any stranger to establish himself on the land. The unsatisfied housing demands of the industrial workers induced many of their colleague who lived in the neighboring town of Tizapán to build and rent rooms for them on their properties, which soon became transformed into full-fledged *vecindades*.

Settlement of the *Ejido* San Bernabé: The Zue Fiction In Cerro Del Judio

Agricultural and industrial employment coexisted in San Bernabé from early on, but the trend of proletarization of the *ejidatarios* intensified greatly after 1940. The *ejido* as an economic unit and way of life based on agriculture began to erode. Groups of *ejidatarios* began to go out to the city to look for work, particularly in the rapidly expanding building trade. Many still managed to combine outside work with cultivation of their parcels. More than forty ejidatarios of San Bernabé found employment in the textile mills of Contreras and Tizapan and at the Loreto paper-mill, while the rest found employment farther away. As their income as salaried workers became more stable and secure, their urban jobs took priority over their agricultural occupations.[10] More and more *ejidatarios* neglected their agricultural parcels as they found city jobs more attractive.

The exhaustion of the communal forest in 1947 further weakened agricultural activity. The *ejidatarios* mismanaged this precious resource. They turned the trees into fire-wood and coal which they sold in the city markets, while the other two villages with rights to the same hills sold their timber as pulp to the paper-mills in neighboring San Angel and Tlalpan.

The first pressures to convert some *ejido* land into urban parcels came in 1940 from within the *ejido* population itself: *ejidatarios* wishing to live near their agricultural parcels, newly married couples lacking housing accommodations and sons of *ejidatarios* living in Mexico City. Although the creation or even the expansion of the *zona urbano ejidal*, requires high level authorization involving a lengthy process, the local *ejido* leadership by itself granted approval to five families to move away from the village to a new site at the foot of Cerro del Judio, thus initiating what became one of the largest irregular settlements in the south of the DF.

The new settlement grew to thirty families by 1950. Among them three families without any ties to the *ejido* which acquired parcels from some of the *ejidatarios*. Such illegal actions were accepted within the *ejidos*, as it was the common practice of the ejidatarios to transfer parcel rights or to rent among themselves. The new site at Cerro del Judio lacked any village services. Electricity was provided from the neighboring Escuela Superior de Guerra, the Higher Military Academy. Water was obtained by buckets from the same source. The *ejidatarios* were also fortunate to be able to board the local buses at the school door.

With the opening of the fashionable Jardines del Pedregal development in the late forties and the building of the UNAM campus in the early fifties, demand for land in the *ejido* began to rise dramatically as local work opportunities increased. Construction workers seeking to establish themselves in the area sought to buy *ejido* land. When demand began to develop in the beginning of the fifties, mayor Uruchurtu issued his development ban (1953). *Ejidatarios* in the new settlement in Cerro del Judio started to sell lots illegally to outside people, reducing their agricultural parcels. The *ejido* Commissariat and the higher agrarian authorities participated directly in these business transactions. At the beginning, sales were made under the cover of including the buyers in the *ejidatario* rosters, or of actually admitting them in the community as members.

The *ejidatarios* began the procedures for recognizing the Cerro del Judio area as a new *zona urbano ejidal* under the 1954 presidential decree of President Ruiz Cortines. This decree mandated that "the property of the lots in the ZUE should be granted free of charge to the recognized *ejido* members, while the excess lots can be sold to individuals who wish to settle within the *ejido* villages to cooperate with the effort to develop them."[11] The *ejidatarios* of San Bernabé decided in a general assembly to request officially the establishment of a *zona urbano ejidal* on Cerro del Judio (July 31, 1955).

The group of unrecognized *ejidatarios* who struggled for official recognition of their *ejido* rights since 1941, feared that they would be excluded from the ZUE under the 1954 decree. They therefore opposed the creation of the ZUE until their membership rights were recognized. The petition was thus delayed for more than four years until April 28, 1959, when it was presented by the original group in spite of the opposition of the other group whose rights were still unrecognized.

The delayed petition, however, had not deterred the *ejido* leaders and the *ejidatarios* from continuing to sell lots. To carry out these informal sale operations, the *ejidatarios* would simply extend to the buyers a handwritten receipt, which they would take to the Commissariat to be "certified" for a fee and a personal reward. It is not difficult to see that in such an anomalous situation of illegal sale in which the local *ejido* authorities played a central role, it was to their interest to maintain a rapid turnover of *ejido* authorities, to dilute responsibilities.

The incorporation of the different ancient villages and towns such as Mixcoac, Coyoacán, Tizapan, San Jerónimo and Contreras on the southern periphery of the city brought about the exodus of their poorer residents, many of whom eventually ended up in such *ejidos* as San Bernabé. As the ancient infrastructure and road-system of these towns was remodeled, sharp rises in housing prices drove poorer residents out to join the growing mass of displaced people looking for housing accommodations farther out in the south of the city.

Demand for urban *ejido* land was also boosted, paradoxically, by the massive loss of textile jobs in the area when the ancient textile mills in the surrounding area closed down between 1955 and 1965. The closing of the outdated textile mills in Contreras and Tizapan left 3,100 workers without jobs. Many workers were dismissed from the Puente Sierra factory due to reorganization. For many workers, job loss meant also losing their company housing accommodations, leaving them to search for new low-cost dwellings in the area.

After thirty five years of petitioning for an access road, an old dream of the *ejidatarios* came true. The order given by the President Adolfo López Matéos, a neighbor of the area, to replace the old dirt road to San Angel which had kept them isolated, with a well-paved modern road, marks the beginning of the end for the *ejido* San Bernabé. Massive demand for land came from outside when the modern access-road to the *ejido* was built (1961-1963) at a time that the Uruchurtu ban on new developments was in full force.

The government itself was instrumental in turning the *ejido* into an irregular settlement, by inducing relocation of people displaced by government projects to the *ejido*. Public works such as highways and urban renewal projects displaced many poor people, who were relocated in the *ejido*. The extension of the Periferico highway in the early sixties enhanced land values along its route and displaced 300 families from Tizapan and from the barracks of the Puente

Sierra textile factory. They were resettled in Cerro del Judio. At the other end of the city, the local government undertook a remodeling project during the sixties in one of the oldest slums, La Candelaria de los Patos, displacing 400 families which were also moved to Cerro del Judio.

As recognition of the rights of the dissident *ejidatario* group was not forthcoming, the conflict between the two groups concerning their ZUE lingered endlessly. Finally, when the new highway to San Bernabé reached the *ejido* in 1963, raising its urban potential and economic value, the deadlock was overcome. A presidential resolution (March, 1963) transferred the membership rights of those *ejidatarios* that had not worked their agricultural parcels for the last two years to 82 *ejidatarios* of the unrecognized group, increasing the number of tenured *ejido* members to 182.

The newly recognized *ejido* members were assigned 82 agricultural parcels for cultivation in Cerro del Judio and inner agreement was finally reached between the two groups in conflict. This land grant, however, was only a cover to allow the opening of the area for urban subdivision and sale, and not to promote agricultural activities. The opposition group was finally coopted and joined the *Comisariado ejidal* in the business of subdividing and selling urban lots. Nevertheless, the official recognition of CJ as a *zona urbano ejidal* was not forthcoming.

Each *ejidatario* on average had at his disposal around one hectare of land (2.47 acres), which he sold after taking care of the needs of his kin. Before selling to strangers, many *ejidatarios* settled their nearest families and more distant relatives. Thus, many extended *ejidatario* families can be found in Cerro del Judio (CJ). *Ejidatarios* not living in CJ, but in the village of San Bernabé, secured their rights by building a house or transferring their lots to relatives. The number of building lots sold and resold began to increase rapidly.

Accusations of illegal sales began to be levelled against the *ejidatarios*. The DAAC, forerunner of the Ministry of the Agrarian Reform, surveyed the situation in 1966 and found that 500 lots had been illegally sold and were occupied by 510 families. Only 198 lot holders among them were San Bernabé *ejidatarios* or long-time residents. The rest had settled there between 1959 and 1966 as a result of illegal acquisition from the local *ejido* authorities. The *ejido* authorities were also accused of sometimes having sold land from the agricultural parcels, without the consent of the ejidatarios. The official DAAC report was made public at an assembly of the *ejido*, and a warning was issued to the *ejidatarios*. The assembly responded by laying the blame on the previous *ejido* authorities.

Two years after the report (1968), the DAAC itself, in a change of face, opened up a special office, and sent in an engineer to certify with his signature and those of the *ejido* president and treasurer the sale transactions. Instead of ending sales, transactions increased greatly. By 1970, at the end of the Díaz Ordaz administration (1964-70), there were already 4,000 lots sold, on which around 40,000 people were living. By 1973 the population had risen to 50,000 residents. The resolution from DAAC recognizing the Cerro del Judio area as a *zona urbano ejidal* was still pending, although in all the official documents the area was considered to be a *zona urbano ejidal.* For the growing number of settlers such a recognition was even more important than for the *ejidatarios,* as the acquisition of their property titles depended on it.

As logically expected, land prices in the informal market increased over time in Cerro del Judio, from a low of $MP 2.50/sq.m. ($US 0.20) in 1963, to $MP 150/sq.m. ($US 12) in 1976, a 60 fold increase in 13 years. It is estimated that up to that point, the sales had brought in the sum of $MP 43.3 million MP ($US 3.46 million) to the *ejidatarios,* which means that each of the 182 *ejidatario* had obtained an estimated income of $MP 237,912 ($US 19,033). Rarely would an *ejidatario* subdivide and sell all his land at once, but would do so in a piecemeal fashion. The payments were often in installments. As a result, the income obtained by the *ejidatarios* was not a lump sum, but was distributed over many years. At the beginning, the *ejidatarios* were inexperienced, and often sold without a clear awareness of prices, but as time went by they became as shrewd as any private owner in their business transactions.

The CJ settlement lacked basic services and infrastructure, legal recognition, property titles and even police surveillance. There was nothing resembling a subdivision plan nor a street plan. An enormous numbers of problems came about, creating endless conflicts between *ejidatarios* and settlers, and within the groups themselves. Family conflicts arose quite often. Many properties became completely locked in, roads were interrupted by fences, fronts of houses found themselves looking to the rear. Those buyers who found that their lots had no access had to negotiate rights-of-way with the neighbors. The agrarian authorities not only became surrogate arbitrators and judges, but planners and urban designers as well, as the DF government followed a strict "hands-off" policy with regard to the *ejido* until 1970.

Regularization of the Cerro Del Judío Settlement

It was not until the inauguration of President Echeverria in 1970, a neighbor from nearby San Jerónimo himself, that the total surrounding area began to receive great attention from the DF government and the highest agrarian authorities, as part of plans to commercialize *ejido* land for high income purposes, to be later explained.

The DF government stepped in for the first time after so many years in CJ, which by then had reached a population of over 40,000 people. As a result of that awakened interest, the authorities of the *delegación*, together with the government party, the PRI, established in 1971 a *Junta de Mejoras Materiales*, a neighborhood council with 15 members with the purpose of improving the area. The council president was also the president of the government party's local chapter. Within two years the *Junta* introduced running water and sewers in the *colonia,* and organized the residents in work brigades to level the streets. The *Junta* thus rapidly climbed to the top of the power-hierarchy of the *colonia* at the side of the *Comisariado ejidal,* the local authority of the *ejido* community. The legal tenure status of the area, however, remained still unresolved.

When the Echeverria administration partially decentralized the DF government, administrative power over the area passed to the local *delegación* of Magdalena Contreras. Notwithstanding that recognition of the *zona urbano ejidal* was still pending, and property titles had not been issued, the Treasury of the DF government applied a property tax on all the lots using the 1971 DAAC register as a basis. The holders received official notification of the tax in 1973, with the provision that taxes were retroactive for five years .

The awakened interest in the area was linked to President Echeverria's plan to capitalize on the economic potential of *ejido* land that could be used for high profit-yielding activities such as tourism or high income development. New tourist centers on *ejido* land were being promoted in different attractive sites throughout the country. These projects were presented to the *ejidatarios* as a practical way of generating new employment opportunities for them in the tourist industry. Presidential expropriation of the land based on agricultural value, and resale by the government to tourism developers based on commercial value, with the profits going to FONAFE, was the central idea behind these projects.

A similar approach was applied to the urbanizing *ejidos* in the urban peripheries. The government approached the issue of tenure regularization of the irregular settlements in terms of the profit yielding potential of the expropriation procedure. Income was to be generated for government by the differential between regularization payments by the settlers and compensation payments to the *ejidatarios.* The economic value of the still vacant lands within the irregular *ejido* settlements was to be tapped through expropriation for the public agrarian sector under the same principle. Logically, expropriation was singled out as the only means of regularization in the irregular settlements to the exclusion of the *zona urbano ejidal* procedure which precluded such income. The regularization procedure was designed as a means of freeing as much land for commercial development purposes as possible and of maximizing the differential between regularization and compensation payments.

Two organizations were designated to implement these plans in San Bernabé, FIDEURBE representing the interests of the DF government, and CORETT those of the agrarian sector. The San Bernabé *ejido* was on the priority list of both of these organizations, due to its high environmental excellence and economic value. On June 8, 1973, only eight days after the creation of FIDEURBE by the DF government, the delegate of Contreras held a press conference in which he presented to the public the redevelopment plan, which proposed the transformation of the Contreras *delegacion* into a tourism center, with artificial lakes, camping grounds and different tourist installations. The plan intended to relocate many of the settlers in compact housing units within CJ, to be built by the *colonos* themselves, and thus to free many hectares not only for tourist installations, but mainly for developing a commercial residential subdivision. This was the first opportunity that the residents had of learning about the new project for their area.

Three weeks after the creation of FIDEURBE, the Mayor of Mexico City Octavio Senties, presented a request to the chief of DAAC, to expropriate in favor of FIDEURBE the 383 hectares of the *ejido* San Bernabé for urban redevelopment. Such a procedure meant that the land would first be transferred in trust to FIDEURBE, and would only then be resold to the settlers. The *ejidatarios* and residents were not previously informed of this decision, which ran counter to the efforts and expectations of the *ejidatarios* and settlers who for 13 years looked forward to the recognition of their irregular settlement as a *zona urbano ejidal*.

The *ejidatarios* were opposed to the FIDEURBE expropriation plan, although it might have produced some income from compensation for them, because they still held more than 100 hectares of vacant land. The settlers on their side also opposed expropriation as it implied paying once again for the land they had already acquired. A few months before the creation of FIDEURBE on March 20, 1973, the *colonos* petitioned the DF government to regularize their tenure rights in the built-up part of CJ, arguing that they acquired their lots in good faith by means of private contracts from the *ejidatarios*, and not as invaders by force. They blamed the *ejidatarios* for having sold them land without considering the necessary spaces for streets.

The *delegación* went ahead very quickly with the implementation of its tourism plan. At the end of July, 1973, the authorities of the *delegación* ordered the eviction of 15 families living on the upper area of CJ, the demolition of their houses, and their transfer to a public housing project in the poorest northeastern quadrant of Mexico city. A violent physical confrontation between the eviction team sent by the *delegación* and local residents stopped the action.

On August 8, 1973, the *Comisariado ejidal* was officially notified about the Major's request to expropriate the *ejido* of San Bernabé. The *Comisariado Ejidal* called together a meeting of *ejidatarios* in which they decided to actively oppose the decree. The redevelopment project and the frustrated eviction attempt undertaken by the *delegación* prompted initially a momentary alliance of *ejidatarios* and settlers against expropriation, although soon enough these two groups parted in different directions because of their differing interests.

A committee to represent the settlers was created by four outside university students with ties to radical groups, who became active among the population in CJ as organizers. The settlers committee was successful in organizing mass demonstrations against the expropriation decree and the redevelopment project. The *Junta de Mejoras Materiales,* in spite of its official ties and the efforts of the *delegación* to co-opt them, also voiced its opposition to the decree. The *Junta* and the settlers committee which meanwhile became the *Consejo de Representantes* (Council of Representatives), jointly requested an appointment with President Echeverria. When their request was turned down, they decided to organize a mass demonstration in front of the presidential palace.

The demonstration of around 800 settlers was organized on November 8, 1973 to the Zócalo, to petition the President to cancel the expropriation procedure. The *ejidatarios,* however, were not informed of this decision, and they did not participate in the mass demonstration. The demonstrators requested the recognition of their settlement as a *zona urbano ejidal,* and the granting of property titles to them as *avecindados.* They thus intended to avoid the intervention of FIDEURBE and the payments required in the case of expropriation. They also requested the introduction of urban services, schools, and transportation by the government, and permission to continue their own building activity, without the intervention of the public-housing agencies. The officials of the DF government and DAAC that were sent out to talk to them rejected the feasibility of the ZUE option and insisted that expropriation and sale through FIDEURBE was the only possible alternative. The leaders decided not to move the people from the Zócalo until they could meet the President in person.

The mass demonstration remained in the Zócalo during the evening, but around 11 P.M. was dissolved by the police, who arrested the participants. With the exception of the leaders, who were kept in jail until the next day, everybody else was quickly released. After this setback, the *Consejo de Representantes* weakened internally, and split into two groups. There were various attempts to organize new mass demonstrations, but each group could only gather small numbers of people. A new organization was formed with splinters of the *Junta de Mejoras Materiales* and the *Consejo* under the leadership of a group of teachers who were trying to establish a school in the area.

The strong opposition from *ejidatarios* and settlers forced the DF government to give up the tourism development project. The residents of San Bernabé, however, were not informed of this change. Some months later (August, 1974) the agrarian authorities informed the residents that the regularization would be carried out through DAAC, which meant that FIDEURBE was not in the picture anymore. Expropriation, nevertheless, would still be the procedure to be followed in the regularization process. They tried to convince the settlers that any other alternative, even if legally possible, such as recognition of the area as *zona urbano ejidal,* would not be in their interests because it would take many years.

Local government party (PRI) leaders then suggested that it would be of benefit to request the assistance of their local Congressman, Rodolfo Echeverria Ruiz, since his intervention could have major effects. After being advised to register as members of the party and being provided with party credentials, a delegation of *ejidatarios* and settlers met with Echeverria Ruiz. They obtained his promise that he would intercede with Mayor Senties to withdraw his expropriation request in favor of FIDEURBE. He requested massive political mobilization of the residents to back him.

The PRI District Committee organized a massive demonstration for the fifth of September 1974 with the backing of the *ejidatarios* and the *Junta*, which was personally attended by Mayor Senties and Congressman Echeverria Ruiz. Congressman Echeverria addressed the Mayor as representative of *ejidatarios* and *colonos*. The Mayor then promised them publicly to withdraw the expropriation request. Ten days later, what seemed impossible happened; the Mayor sent a letter to the DAAC, with a copy to the *ejidatarios*, requesting the withdrawal of his previous expropriation petition, and clearly stating the convenience of achieving privatization via the "agrarian legislation procedure," whatever that meant.

The withdrawal of the expropriation request was favorably received by the *ejidatarios.* It was in their interest to postpone as much as possible the process of expropriation, considering that they still held much unsold *ejido* land which had gained substantially in market value. With this move, the Junta and the PRI momentarily outmaneuvered the *Consejo*, which found its flag wrested from its hands. However, barely 15 days later (October 30, 1974), Mayor Senties reversed himself and sent a letter to the head of DAAC stating: "After better analysis of the case, a conclusion has been reached that the conditions which motivated my original request of expropriation are once again valid, for which reason I request that the formalities of the expropriation process continue as foreseen by the law."

On November 8, 1974, a new charter restructuring CORETT was issued, by which it was authorized to promote expropriations of *ejido* land to regularize property titles. In January 1975, CORETT requested from the DAAC that if expropriation be decided, the decree should be issued in its favor instead of FIDEURBE, but should not cover the total 383 hectares intended by FIDEURBE but only 260, seemingly to assuage *ejidatario* opposition. DAAC, which was meanwhile raised to the rank of a ministry, undertook a census of the parcels in the *ejido* and an assessment of their value to determine the payments required to legalize the lot titles. They received the cooperation of the *Junta de Mejoras Materiales*. Active opposition came forth from the *Consejo de Representantes* which once again raised the flag of legalization via the ZUE procedure.

The census showed there were 5,345 lots in Cerro del Judio covering 163 hectares. These were divided into five zones for assessment purposes. Their assessed value varied between $5.00 MP/sq.m. ($US 0.40) and $MP 15.00/sq.m. ($US 1.20). A total payment by the residents to CORETT of $MP 14,335,419 ($US 1,146,833) for the 163 hectares of land was determined. The official decree expropriating the *ejido* San Bernabé in favor of CORETT, was finally issued in August, 1975.[12] It stated that the *ejidatarios* would get their property titles free of charge.

To start the process of regularization CORETT called a general assembly on September 20, 1975. Five thousand settlers and *ejidatarios* came to a meeting with the general director of CORETT. Although he was sure of their acceptance in view of the low assessment value, he encountered fierce opposition from the *ejidatarios* and a vocal minority of settlers. The *ejidatarios* argued that CORETT was getting the lion's share from the regularization payments, $MP 8.00/square meter ($US 0.64), while they would only be receiving $MP 1.50/square meter ($US 0.12). The *Consejo de Representantes* also opposed

CORETT, and some *colonos* reminded the director of CORETT about the Mayor's broken promise. When tension mounted, the director broke off the stormy meeting and left under heavy guard.

CORETT, nevertheless, was able to implement its regularization procedure, as the overwhelming majority of the settlers was won over by the low assessment value. After the stormy assembly, CORETT, in retaliation gave up the idea of opening up a local branch , and individually summoned each lot holder to its central offices. Within one year, seventy percent of the settlers came forward to sign their individual agreements with CORETT. CORETT, in this case, was able to achieve regularization by winning over to their side the settlers who far outnumbered the *ejidatarios.*

The *ejidatarios* on their part went to court and asked for an *amparo,* a court injunction against CORETT, which was granted them (December 18, 1975). The higher level court upheld the *ejidatarios* appeal (February, 1979) and ruled that a new assessment was required considering the rising values of the surrounding area which had undergone major development. As a result CORETT was legally unable to deliver the property titles it had promised to the settlers.[13]

CORETT again filed an official request (February 26, 1979) to the Ministry of the Agrarian Reform to expropriate the total 389.4 hectares of land belonging to the San Bernabé *ejido.* CORETT's request was approved two years later and the expropriatory decree was published in the Diario Oficial.[14] To win time and continue the land sales the *ejidatarios* once again challenged the expropriatory decree and appealed to the federal courts against CORETT. The illegal sales of land by the *ejidatarios* continued and the population in CJ increased so rapidly that by 1980 it contained around seventy thousand residents in six *colonias* which covered the total area of the *ejido* de San Bernabé.

Four years later the court granted the *ejidatarios'* appeal (June 18, 1985). As a result CORETT was forced to renegotiate its compensation payments with the *ejido,* and after another three years (June 18, 1988) arrived at a new agreement with the *ejidatarios.* This agreement was turned into a new expropriatory decree recently signed by President Salinas de Gortari.[15]

According to the new agreement the area expropriated covered 342 hectares. The compensation payments were assessed on the basis of $MP 150/sq. m. ($US 0.06/sq. m.), making a total of $MP 513,013,616 ($US 205,205). Under the assumption that the number of recognized *ejido* members still stood around 182, the compensation to each ejidatario amounted to $MP 2,818,898 ($US 1,127.5). This was the compensation that the ejidatarios achieved after a struggle of more than sixteen years. Half of the sum was to be deposited in the *ejido* account at FIFONAFE within thirty days, while the other half would become available according to the cash flow of the regularization process. The decree required that the funds to be deposited in FIFONAFE were subject to the guidelines of the Agrarian Reform Law,[16] which establishes that only when the investment plan approved by the Ministry of the Agrarian Reform is not carried out by FIFONAFE within one year of the deposit, can a cash distribution of the money between the *ejidatarios* take place.

CASE STUDY 2: THE COLONIA POPULAR MIGUÉL HIDALGO

This case study by Lugo Medina & Bejarano González,[17] surveys the development of the irregular settlement Colonia Popular Miguél Hidalgo which was created on the land of the *ejido* Tlalpan. On the rocky lava formations of the former *ejido* now stand beside *Colonia Popular* Miguel Hidalgo, the grand campus of the Mexican National University (UNAM) and the extension of Jardines del Pedregal de San Angel, one of the most exclusive residential developments in the country. The Colonia Popular Miguél Hidalgo began in the early fifties as the *zona urbano ejidal* of the *ejido* on a site known as Rancho del Teochihuitl.

Historical Origins Of *Ejido* Tlalpan

The *ejido* Tlalpan is located within the *delegación* of Tlalpan to the south of Mexico City. Various free Indian villages surrounded by a number of *haciendas* and *ranchos* existed since colonial times in what is now the *delegación* of Tlalpan such as Huipulco, San Pedro Mártir, San Andrés Totoltepec and San Agustín de las Cuevas, the later day Villa de Tlalpan. Tlalpan also became the site of textile factories during the Juárez period (1856-72). Some paper mills were drawn to the area in the second half of the nineteenth century by the abundant supply of raw materials in the nearby forests of the Sierra del Ajusco, Contreras, and Milpa Alta and the water resources from the Tlalpan river and the Fuentes Brotantes.[18]

Property rights problems led to conflicts and litigation between the villages and the *haciendas* and later between them and the factories.[19] They became involved in conflicts over water rights with the factories in the late nineteenth century.[20] After years of unsuccessful litigation, the peasants of the various villages of Tlalpan, as many other thousands of peasants in the south, joined Zapata in the armed struggle that began in 1910, demanding—under the banner of the Plan de Ayala—the return of the lands, forests, and waters, that they claimed were taken away by the *haciendas* and factories.

The peasants of the Poblado of Tlalpan were one of the earliest claimants of land restitution from the Carranza government in 1916. Their claim of restitution was not recognized by the revolutionary government, but they were nonetheless considered entitled to an endowment of *ejido* land according to the new Constitution. The revolutionary government designated land from the Rancho del Teochihuitl and the Hacienda Peña Pobre for the endowment. These estates were owned by the La Fama Montañesa textile plant and the Peña Pobre paper mill which turned to the courts for an injunction, delaying the endowment process for many years. President Emilio Portes Gil (1928-30) finally signed the endowment decree which established the *ejido* Tlalpan in 1930,[21] while recognizing at the same time the factory's right to compensation and to part of the water resources.

The 1930 endowment decree granted 985 hectares of land to 330 peasants to establish the *ejido* Tlalpan. 525 has. of land were taken for that purpose from theRáncho de Teochihuitl and 460 has. from the Peña Pobre *hacienda*.. The *ejido* Tlalpan petitioned again during the Cárdenas period (1934-40) for an enlargement of their endowment for a growing number of new applicants. President Cárdenas granted 426 additional hectares which were taken from the Hacienda de Copilco to 211 of the applicants, while 205 applicants were still left on the waiting list.[22] The enlargement area was not adjacent to the original endowment area. Only a small fraction of the land granted to the *ejido*, not more than 8%, was good for agriculture, while most of it belonged to the so called *agostadero* class, a type of land only fit for animal grazing and for the cultivation of certain fruit tree species.

Disappearance of the *Ejido* Tlalpan: Conversion of its Land

The urbanization process which led to the disappearance of the *ejido* Tlalpan, as of many other *ejidos* in the southern periphery, began in the forties,[23] not more than a decade after the *ejidatarios* obtained their first endowment. The government of President Alemán (1946-52) chose the area to establish Ciudad Universitaria, the campus of the National University and expropriated part of the *ejido* for public interest purposes. The rest of the *ejido* was legally bartered under government surveillance with private owners, while a minor fraction was designated as a *zona urbano ejidal.*

The first expropriation of land from the *ejido* Tlalpan occurred in 1944, when 4.3 Ha. were taken for the construction of the Mexico-Acapulco highway.[24] This was followed in 1946 by the expropriation of 733 has., almost half of the total area of the *ejido,* to build Ciudad Universitaria, the campus of the National University of Mexico (UNAM).[25] The monetary compensation paid to the *ejidatarios* for this land totalled the amount of $MP 95,270, ($US 19,643), less than one Mexican cent per square meter. Another 4.3 hectares were again expropriated from the Tlalpan *ejido* in 1954 for the Mexico-Acapulco highway for $MP 6,540 ($US 523).[26]

In addition to expropriation, a major barter operation involving land belonging to the *ejido* was also carried out with private owners on the basis of article 146 of the old Codigo Agrario. An area of 795 hectares of the *ejido* of Tlalpan was exchanged in 1959 for two agricultural properties of equivalent size in the state of Guanajuato.[27] The bartered portion of the Tlalpan *ejido* was used to expand the upper-class development Jardines del Pedregal de San Angel.[28]

Due to the higher assessed value of the Tlalpan land, the *ejidatarios* of Tlalpan received an additional monetary compensation of $MP 8,521,964 ($US 681,757) in the barter operation. The money was deposited in the Banco Ejidal and later transferred to FONAFE[29] for investment purposes. From a purely monetary basis it is clear that the *ejidatarios* from the Tlalpan *ejido* struck a far better deal through barter with the private owners than through expropriation with the government. For half of their total land which was expropriated by the government for UNAM, they received the sum of $US 19,693, while from the private barter operation they obtained, in addition to the full replacement of their land in Guanajuato, the sum of $US 681,757 which was deposited in their FONAFE account.[30] However, the *ejidatarios* were seemingly unable to retrieve part of that money, at least not until 1980.

As a result of the barter operation, 57 of the 236 *ejidatarios* decided to resettle in Guanajuato to farm their land, while the majority stayed behind in Tlalpan to develop the *zona urbano ejidal,* which was officially transferred to a new site when the barter operation was approved. With the barter operation, *ejido* Tlalpan ceased to exist for all practical purposes as an agricultural entity in the DF. The interests of the two *ejidatario* groups separated, and conflicts emerged between them, although they still shared in common many interests vis-a-vis the settlers as the legal owners of the ZUE in the *colonia* .

The actual division of the *ejido* community into two different groups, one staying behind in the Tlalpan ZUE and the other moving to their new location in Guanajuato, was decided six years after the publication of the barter decree, in a general assembly on August 8, 1965. There they approved the use of $MP 2,000,000 MP ($US 160,000) of their deposited funds at FONAFE for the development of a tourism project in Guanajuato suggested by FONAFE and another $MP 2,000,000 ($US 160,000) for Tlalpan to create small agricultural farms which would allow those that stayed behind to maintain in force their rights as *ejidatarios*.

The Guanajuato project included the building of an access road, an artificial lake, swimming pools and handball courts on part of their land. A special committee of *ejidatarios* was appointed to supervise the project in Guanajuato. Only a few months transpired, and the *ejidatario* in Guanajuato became disillusioned with the tourism project and drew the conclusion that it was nothing but a scam perpetrated by the FONAFE officials, in connivance with some of their own leadership. As a result, forty five *ejidatarios* journeyed back from Guanajuato to Tlalpan, and in November 1966 demanded the dismissal of FONAFE and the return of the money. The reprisals did not delay much in coming, and the Guanajuato *ejidatarios* were soon denied the yearly credits for seeds and fertilizers by the Banco Ejidal.

The struggle of the Guanajuato *ejidatarios* against FONAFE continued. Two years later in 1968 at the height of political turmoil in the country they published a series of public letters in the press, denouncing the barter operation as a fraud. They claimed that theirs was not an isolated case, but that FONAFE had defrauded a minimum of 10 other *ejidos* in the Federal District by means of barter transactions. They also brought to public attention that their land in Guanajuato was being invaded by other peasants applying for land, and that they were being subject to unfair practices in the distribution of water.

Settlement of Miguel Hidalgo

The origin of the Colonia Popular Miguél Hidalgo as an irregular settlement goes back to a request that the *ejidatarios* of the *ejido* Tlalpan addressed in 1946 to the agrarian authorities to establish a *zona urbano ejidal*, after the larger part of their *ejido* was expropriated for the UNAM campus. The request was granted in a resolution signed by President Miguél Alemán in 1951.[31] An area of 169.81 hectare located between the future Ciudad Universitaria and the Peña Pobre factory was assigned for that purpose. A total of 236 lots were to be subdivided for the *ejidatarios* and for a small number of *avecindados* .

The ZUE was actually not developed at the isolated site designated in the original decree. It was transferred instead by the *ejidatarios* themselves before obtaining legal approval, to another area within the *ejido*, known as El Teochihuitl, in the vicinity of the textile plants and their workers quarters, which were already being irregularly settled. The new site was more attractive to the *ejidatarios* because of the existence of the nearby factories and private developments. The rocky lava formation of the area was most inappropriate for low income settlement due to the difficulties and great cost of supplying services, sewers in particular. Such land was appropriate for very low density developments like the UNAM campus or Jardines del Pedregal, but totally inadequate for high-density, low-income settlement.

The new settlement adopted the name Colonia Popular Miguél Hidalgo. Eight years passed until the transfer of the ZUE to the Teochihuitl site was granted legal recognition in 1959. The decree which approved the barter operation in 1959 also relocated the ZUE to the Teochihuitl area when it already contained a population of three thousand irregular residents.[32] The *ejidatarios* were assigned 236 plots of 5,000 sq. m. each in the relocated ZUE which they very soon began to sell.

Increase of demand for low income housing in the area came as a result of new employment opportunities that became available in the south of the DF in the fifties and sixties at a time that the Uruchurtu development ban was in force. The expansion of the existing local industries in the vicinity, various major construction projects including the Villa Olimpica for the World Olympic games, the appearance of new industrial, commercial, and building investments in the nearby private developments, all increased demand. These expanded employment opportunities often brought in poor migrants from other states who joined their relatives and friends in the Miguél Hidalgo settlement. Demand also grew as a result of the area's increased attraction for people working in other parts of the city, due to the great improvements in communications and transportation that took place in the region by the extension of the Anillo Periferico and Insurgentes highways, and later by the building of the Picacho-Ajusco highway in connection with the Olympic games.

Commercial subdivision opportunities were discerned by the local *ejido* authorities very early on. They were already selling informally unserviced lots to settlers in the early fifties in the Teochihuitl site, long before it was recognized as a *zona urbano ejidal.* After the 1959 barter decree that officialy transfered their ZUE to the Teochihuitl site, the population began to grow rapidly there. From 1960 to 1970 the population increased from 3,447 to 14,414 residents, more than 4 times its former number, and even more rapidly in the next decade.

There are three distinct zones in the *colonia* which correspond to the different development periods. The lower part of the *colonia*, the oldest, was settled between 1950 and 1967 and developed the highest service level and densities; the extension *(ampliación)* of the colonia, was settled between the years of 1967 and 1972 reaching lower densities and level of services; the upper section of the *colonia* is the most recent and was opened without services at all after the Picacho-Ajusco panoramic highway passing through was built in 1972. The residents of the three sections acted more often than not in an independent and even contradictory manner in their negotiations with government agencies, reflecting their differences rather than their common interests.

There were various means by which acquisition of lots took place in Miguel Hidalgo. Money was one of them, physical work another one, and political influence still another one. The *ejidatarios* subdivided and sold small parcels directly or through the *comisariado ejidal* to interested buyers for cash. These buyers often resold or transfered the lots acquired from the *ejidatarios* or the *comisariado ejidal* to other private settlers. The comisariado *ejidal* was also involved in direct sales to speculative buyers acquiring lots for profit. Many of these buyers who possessed one or various lots in the *colonia* would seek to protect their property against invaders by installing a watchmen or a free tenant on the lot.

Some settlers acquired lots in exchange for physical work in the opening of streets and leveling the land in the new sections, and the payment of weekly monetary quotas; after opening and subdividing a new area, the participants of the work teams would take part in a lottery by means of which the distribution of lots would take place.[33] These lot holders would receive some unofficial documentary proof from the *comisariado* to support their claim to the property.

Another method involved the assignment of lots to privileged recipients of the agrarian bureaucracy. Officials and employees of the DAAC or of the Ministry of the Agrarian Reform, were assigned lots free of cost by the *delegado agrario* and the *comisariado ejidal* at various periods.[34] These lots, which were generally not intended to be personally settled by the recipients, were put under the custody of watchmen, to wait for their increased value as a result of legalization and the introduction of services.

There were also squatters within the ZUE in the areas known as El Metro, El Capulín and Cartolandía. The squatters occupied ZUE lots that some *ejidatarios* had previously sold to the promoters of the barter operation for $MP 55,000 ($US 4,400) each.[35] The promoters resold these lots to ISSSTE, the social security agency of the federal bureaucracy. A fraction of the ISSSTE property was invaded by squatters. CORETT refused to regularize tenure to them when the regularization program was implemented. Squatters also gained a foothold in the upper part of the colonia after the implementation of the regularization program.

The irregular situation of the *colonia* provided the city authorities with an excuse to deny services to the settlement for 25 years. Electricity, which is generally introduced in the irregular settlements very early on regardless of their status, came as late as 1965. At some point in time the *delegación* closed off the area to the *peseros*, the informal collective taxi cabs linking the *colonia* to the external world, making life particularly miserable for the settlers. Building permits were universally denied, and thus all the dwellings were built outside of the law. A police booth was installed at the entrance to the *colonia* by the *delegación* authorities, forbidding access to any vehicle carrying building materials. The settlers, however, found a practical although costly way to go around this problem by bribing the police every time such vehicles entered the *colonia* .

Water was carried in buckets by women and children from as far as the Fuentes Brotantes for many years. Later on, under strong pressure from the settlers, the *delegación* began to supply them with a weekly ration of one drum of water per family, which was distributed by a tanker truck. The irregularity and unreliability of the service made it necessary for the settlers to buy additional water from private tankers. It was not until CORETT began to operate in the area in 1975 that the first hydrants were installed on street corners in the lower part of the *colonia,* to which the settlers contributed their work, while the rest continued to rely on tanker service.

A survey by CORETT in the early seventies, 25 years after the beginning of the *colonia,* revealed that from the total number of 1,274 dwellings existing in the settlement, over 99.2% lacked sewers, 89.9% lacked running water, 15% were not connected to the electric power network, while 21.2% were illegally hooked up to it. 66.5% of the dwellings had roofs of provisional materials like cardboard, asbestos and metal sheet, 48.8% had only one room, while 62.6% used the kitchen as a sleeping space.

The oldest settler organization in the *colonia* was the *Junta de Mejoramiento Moral Cívico-Material,* which was organized in the early years by the lower section residents to petition for services and tenure regularization. Internal friction developed in the *Junta* in the late sixties when the presidency was taken over by an external representative of the PRI, a female school teacher who had the mission of coopting the *Junta* members into the government party. The opposition members broke away from the *Junta* in 1971 and created the *Asociación Civil,* whose main objective was similar to that of the *Junta,* such as petitioning the city government and the Departamento Agrario (DAAC) for the regularization of their individual ownership titles and the introduction of services. The *Asociación*, which at the highest point of its popularity had around 200 members, was substantially weakened when one of its leaders—an ex nun—defected from their ranks and joined the PRI.

In 1974 a group of settlers in the extension of the *colonia*—the then most disadvantaged area in its lack of services—created a new *Asociación Civil* which, like that of the lower section, adopted as its main goal the regularization of land ownership and the introduction of services. The Asociación became actively involved in various confrontations with the city government. It fought the police ban on the entrance of building materials and the closing of the area to the *pesero* taxi cabs. It was also instrumental in organizing a property tax payments strike against the DF treasury in 1975 when the DF government imposed property taxes in 1961 on them on the basis of a ruling by the *Procuraduria de Colonias* that all the settlements had to pay taxes even when they were on *ejido* land. When property assessment values were raised in the area in 1975, after regularization was being implemented, the Asociación again organized a tax strike.

A split developed, or was rather stirred up within the ranks of the *ejidatarios* and between them and the settlers in the beginning of the seventies. The issue was the control of the *comisariado ejidal*—the executive leadership of the *ejido*—in charge of the informal subdivision and sale of the ZUE in Tlalpan to *avecindados*. When new elections were held in Tlalpan for new officers of the *comisariado ejidal* on June 27, 1971, the new *comisariado* reassigned or resold some of the lots that the *delegado agrario*, representative of DAAC in the *ejido*, and the previous *comisariado ejidal* had assigned for free in the ZUE to officials of the agrarian bureaucracy.

The 1971 elections were soon after questioned as invalid by a coalition of the Guanajuato group and a group of Tlalpan *ejidatarios* supported by the CNC. DAAC, then the highest agrarian authority, backed them and invalidated on September, 1972 the 1971 elections, on grounds that only 14 of the 88 *ejidatarios* who voted in the assembly had voting rights. The DAAC decision was based on the premise that only those *ejidatarios* who moved to Guanajuato and engaged in agriculture were still legitimate *ejidatarios*.

Soon after the elections were invalidated, the rumor spread in the *colonia* that a large group of invaders incited by the *comisariado* was ready to take over and displace the settlers. On the night of September 15, 1972, the eve of Independence Day, the settlers stood on guard with firearms to prevent such an invasion. In the morning a full police detachment appeared in the *colonia* accusing the settlers of being invaders. After removing the settlers' guards, demolishing various dwellings, and beating up some residents, they finally left.[36]

The next day, the settlers, led by a natural leader, a textile worker named Vicente Guerrero, staged a mass demonstration with placards to the zócalo to protest the police action. The *delegado agrario*, representative of the Ministry of the Agrarian Reform in the *ejido*, strongly objected to the settlers' mobilization. A delegation of five settlers, led by the textile worker, was given access to see President Echeverria. After the President contacted mayor Octavio Sentíes, they were sent to the *Subprocuradoria de Colonias* .

An agreement was reached there which granted recognition to the settler committee to take over from the *comisariado ejidal* the control of the documentation of lot distribution, and the organization of the settlers collective work tasks. This obviously wrested away from the acting *comisariado* some of its basic functions. The settlers, however, did not capitalize on their apparent triumph. They claim that their leader was threatened with death and forced to leave the *colonia*. Without its leader the settlers' movement collapsed, and the *delegado agrario* and the comisariado *ejidal* finally regained total control over the subdivision and commercialization process of the ZUE in 1973. Nevertheless no new elections for the *comisariado* were held during the Echeverria administration until 1977, when control of the *comisariado* was gained by the Tlalpan opposition group. In the meantime the deposed *comisariado* continued in its functions selling and transfering lots.

Regularization of Miguel Hidalgo

CORETT, the intergovernmental commission to regularize illegally occupied *ejido* property began to operate nationally in 1974, and at the end of that year stepped into the *colonia* Miguél Hidalgo. The residents were informed by CORETT that the *colonia* would have to be regularized through expropriation, and not as a *zona urbano ejidal.* Thus, on December 31, 1974, an official request was presented by CORETT to the newly created Ministry of the Agrarian Reform to expropriate the 169.81 Ha. of the *colonia* for regularization. The decree expropriating the land in favor of CORETT was issued by the Ministry of the Agrarian Reform a year later on January 2, 1976. It authorized CORETT to take over any existing vacant lots for commercial purposes. The profits of the operation—the difference between the price charged to the settlers and the compensation payments—would be divided up by CORETT in 2 parts: 20% for the *ejidatarios* and 80% for FONAFE and CORETT.

A socio-economic survey of CORETT in 1974 showed a population of around 15,000 people between settlers and *ejidatarios*. There were a total of 1271 occupied lots in the *colonia* that had to be regularized. Agricultural activities had practically ceased among the population, with less than one percent engaged in agriculture. Twenty five percent of the total population was employed: 34.4% in the services sector, 27% were blue collar workers, 20.8% were white collar employees, 5.7% businessmen and 2.6% professionals. Among the heads of families, 55% had steady employment, 41% held unsteady jobs, and 4% did not work at all. At a time that the official minimum monthly wage was set at $1,902 MP ($152.2 US), 32% of the family heads earned above above that limit, 40.5% earned between $1,500 and $2,000 (between$120 and $150 US) and 27.5% earned below MP$1500.

Settler Response to the Regularization Program

The interests of the settlers as a group came in conflict with those of the *ejidatarios*. The clash of interests came into sharp focus when the regularization programs were implemented and CORETT entered the settlement to regularize the tenure situation. Settler reaction to the expropriation decree and to CORETT was different to that of the *ejidatarios*. While private property titles were a long cherished goal shared by *ejidatarios* and settlers alike, the *ejidatarios* were seeking to increase their share in the proceeds of the operation, while the settlers were interested in minimizing payments. These two groups did not join together in their struggle with CORETT, and were mostly at odds with each other.

Notwithstanding that regularization by CORETT looked like paying twice for the same goods, the majority of the settlers soon submitted themselves to the CORETT regularization program. The fact that the commercial value of their lots was far above the price set by CORETT, and that these payments would protect them from eviction, helping them to achieve their long cherished goal of property titles and public services, explains why the efforts to stir up the settlers against regularization was largely unsuccessful.

CORETT's strategy in the *colonia* was carefully laid out. To determine the regularization payments that the settlers were required to make to CORETT, the *colonia* was divided into 4 distinct zones whose payments varied according to their proximity to the Insurgentes and Picacho-Ajusco highways. They varied from a minimum of $MP 15/sq. m. ($US 1.2/sq. m.) to a maximum of $MP 45/sq. m. ($US 3.6/sq. m.). Such payments, which were from five to fifteen times the agricultural value, were still substantially below commercial value. They were to be paid by the settlers in cash or on easy low-interest installments. In addition to these payments, the settlers were required to cover the ten percent tax on transfer of property.

By dividing the *colonia* into four zones with different prices, CORETT made sure that the basic reason that could have united the 15,000 settlers against its intervention would be minimized. In spite of some issues that could have united the settlers against CORETT, such as its refusal to recognize the legitimacy of the settlers in the upper section, CORETT was successful in not allowing these problems to come to the fore in the beginning. To win the acceptance of the settlers for the regularization program, and to induce them to come forth individually to sign their agreements, CORETT relied for its public relations on the human networks of the national child-protection agency INPI and the government party (PRI). INPI, the government organization for child care and family services, was quite popular in the *colonia*, and the PRI, the government party, had active followers in the various settler organizations.

The regularization program brought to life a very active independent settler group. The independent group was led by a charismatic Claretian priest, a follower of the so-called Liberation Theology, who had come to the *colonia* in 1972 and had close contact with the settlers through the parish church in the lower section of the *colonia* and the chapel in the extension section. He was successful in drawing to him various youth groups in order to create an independent workers-settlers movement, which would join in the struggle for general political and economic change. The group began an intensive organizing activity around the regularization issue aimed at unifying the settlers of the different sections of the colonia around a common platform. Calling in the more militant members of the settler organizations, the group organized meetings in the different sections of the *colonia* to draw up a petition sheet to CORETT.

In the first round of meetings in January 1976 they put together a nine-point sheet which contained such demands as a public discussion of the contract clauses with CORETT; a minimum payment period of eight-to-ten years ; elimination of down payments, deed registration fees, and cancellation clauses due to delayed payments; maintenance of the present lot size and non-commercial prices for vacant lots; and incorporation of the *colonia* as a popular settlement area to forestall any future tax increases. The independent group called a second round of meetings in February 1976 to elect block and section representatives. Once these representatives were elected, the final petition sheet to CORETT was drawn up with help from the other organizations in the *colonia* and compressed into four points: 1) fair land prices in accordance with the settlers' economic possibilities; 2) Collective bargaining before accepting any decision; 3) incorporation of the *colonia* as a low-income settlement to forestall any tax increase; 4) recognition of the possession rights of the present residents, regardless of how these where obtained.

Still unsecure about their power-base and demands the demand sheet was not yet presented to CORETT. Internal divisions arose within the group, and its energy was spent in discussions on organizational structure. In spite of the turmoil it created, the independent group declined rapidly. The priest was accused by some of his followers of paternalistic leadership. Most of the representatives soon ceased to attend its meetings. The independent movement practically ceased to exist when the priest was suddenly ordered by the superiors of his Order to leave the *colonia* at the end of the year (1976), after a four-year stay there.

The regularization program finally began to win acceptance among the settlers and their organizations. The various settler organizations which had advocated property titles and the introduction of services were finally led to passive consent. The majority of the members in the *Asociación Civil* of the lower section of the *colonia* accepted the leadership of the ex nun who defended the decree against the attacks of the *ejidatario* opposition group and the independent settlers' group. Her viewpoint that the decree represented the best deal they could get prevailed against the dissenting voices of two other smaller factions within the *Asociación.*

One dissenting view was advocated by a lawyer, himself a settler in the lower section, who thought that they could side step the payments and obtain recognition of their property rights by applying the legal principle of positive prescription which brings about recognition of ownership after a certain number of years in cases where peaceful unobstructed occupation of a site has occurred. His view, however, was eventually rejected when it became clear that the Constitution excludes positive prescription from taking place in the case of *ejido* property. Another faction in the *Asociación* of the lower section which had participated in the tax payments strike, was in favor of joining the independent group in active opposition to CORETT. As a whole, however, the *Asociación* was neutralized by internal dissension and weakened in its militancy by the well-timed introduction of some services by the Tlalpan *delegación* in the lower part of the *colonia* in 1975.

The *Asociación Civil* of the extension section showed apprehension towards the regularization program at the beginning, and forwarded a petition to the director of CORETT, in which they presented a list demanding recognition of the rights of any settler with more than 2 years of continuous uninterrupted occupation, regardless of the way he took possession. They also demanded adjustment of the monthly payments to the particular payment capacity of each family, and a minimum payment period of 5 years without any down payment. Although CORETT never responded to these demands, the *Asociación* of the extension section soon gave up active opposition to CORETT when it learned that the extension section had been appraised at the lowest value of the *colonia* , $MP 17/sq. m., due to its minimal services. By the end of 1976 most of its members had signed their contracts with CORETT.

When the agreement with the *ejidatarios* was reached in July 1976, CORETT established a branch office in the *colonia* and began the regularization process in August. With the aid of the *ejidatarios*, INPI and the PRI, a successful campaign was launched by CORETT to induce the settlers to sign their contracts. The more wealthy, the non-resident, and the owners of various lots in the colonia were the first to sign up, most of them paying the total sum in cash. The settler organizations were still divided, but many of their followers came forward to sign their contracts. Thus in spite of the activities of the independent settler group and the *ejidatario* opposition group, the overwhelming majority of the settlers seized the opportunity to regularize their property titles when CORETT opened its offices in the *colonia*. In November 1976, CORETT handed over the first 100 property titles to the settlers.

Ejidatario Response to the Regularization Program

The original expropriation decree was drawn up in a way that favored the Guanajuato group creating a split among the *ejidatarios*. The decree applied a clause of the agrarian reform law (article 123), hardly applicable to this case, which states that where **part** of an *ejido* is expropriated, the compensation sum should be used to replace the land expropriated with other agricultural land or to promote agricultural development in the *ejido*. This meant that the compensation payments were to be placed at the disposal of FONAFE to acquire land to expand the *ejido* in Guanajuato or to carry out an agricultural development program in Guanajuato.

The *ejidatarios* split into three groups as a result of the expropriation decree. In one group were the Guanajuato ejidatarios who claimed greater rights than the rest of the *ejidatarios*. In the second group was a minority of Tlalpan *ejidatarios* holding control of the *comisariado ejidal* who agreed to the expropriation decree without consulting the *ejido* assembly. A third group which held the majority of Tlalpan *ejidatarios* opposed the decree due to its exclusionary terms and began an active struggle against the terms of the regularization operation.[37] This group objected to the compensation amounts and to their exclusion from the subdivision and commercialization process.

Compensation payment due to the *ejidatarios*, was assessed on twice the agricultural value of the land at $MP 6/sq. m. ($MP 0.24/sq. m.), although its commercial value fluctuated at that moment in the open market between $MP 250 and $MP 500 MP, 125 times as much on average![38] Global compensation totalled the amount of $10,189,080 MP ($815,126 US). The compensation payments were to be divided between CORETT which would receive 80% of the total and the *ejidatarios* receiving the remaining 20%. Their share would be credited according to the cash flow of the regularization process which could extend over a period of years. The decree expropriating the land in favor of CORETT also authorized CORETT to take over any existing vacant lots for commercial purposes.

The Tlalpan majority group became the leading opposition force to confront CORETT. The conflict continued for various years in different forms, being basically concerned with obtaining a fair share in the proceeds of the land sales. A month after the expropriation decree was issued (February, 1976) the Tlalpan group threatened to stop the regularization process through a court injunction unless their demands were met. They requested a higher land assessment value, direct payment of the compensation sums to the *ejidatarios* without the intervention of FONAFE, participation of the *ejidatarios* in the process of lot assignment, commercialization of the vacant areas and special contracting terms for the descendents of the *ejidatarios*.

The Tlalpan opposition group mobilized eleven other *ejido* communities in the DF through the League of Agrarian Communities of the DF which were part of the CNC, the peasant block within the PRI. To pressure CORETT to change its stand, five thousand *ejidatarios* from these *ejidos* marched on March 22, 1976 to the *colonia* Miguél Hidalgo in solidarity with the Tlalpan *ejidatarios*. They achieved the dismissal of the agrarian delegate and signed an agreement with CORETT which met many of the demands of the Tlalpan opposition group on July 22, 1976.

The Tlalpan *ejidatarios* were successful in increasing the initial compensation by another $MP 4 /sq. m. making a guaranteed minimum of $MP 10/sq. m. to be deduced from the 20% benefits due to the *ejidatarios*. They also achieved that the compensation money should be spent on housing improvements mainly benefitting the Tlalpan *ejidatarios* instead of being invested in Guanajuato.

CORETT also agreed tactically to a list of demands by the ejidatarios which included the following points: that FONAFE be eliminated as executor of the improvement program, allowing other government organizations to carry out the program; that CORETT should give recognition to the lot assignments that the *ejidatarios* would determine in their assembly and give priority to those candidates that the *ejido* assembly would recommend for vacant lots or to take the place of those who would not come forward within six months to sign their regularization contracts; that one lot per *ejidatario* family be assessed at the lowest regularization price for the descendants of the 235 *ejidatario* families, regardless of location. CORETT also agreed to assist the *ejidatarios* in petitioning the then Secretary of the Agrarian Reform, Felix Barra, to return them the missing $MP 4,400,000 ($US 352,000) which were deposited 17 years before in FONAFE, when the barter operation had been carried out.

An important point in the agreement concerned the question of who would assign, commercialize, and benefit from the still vacant lots in the upper part of the *colonia* which was opened up by the construction of the Picacho-Ajusco highway. CORETT, seeking to increase its benefits from the regularization programs, was trying to expropriate as many vacant areas as possible to sell them at their full commercial value, or to assign them to designated political beneficiaries. However, to reach an agreement with the *ejidatarios*, it gave them assurances that the proceeds of the lots in the upper section of the *colonia* would be used to their benefit to establish *ejido* enterprises. CORETT also promised them that the compensation due for the land which was taken to build the highway would in its entirety go to them. These promises later became a source of strife when CORETT did not comply with them.

When CORETT began to sell 250 vacant lots at the end of 1976 in the upper section on both sides of the Picacho highway breaking its earlier agreement of July, a conflict broke out. The whole section contained a total of 22 blocks which were highly valued due to their commercial potential, although they lacked almost all the services with the exception of an electric power line to which the settlers were hooked up illegally. Half of the blocks in the upper section were still unoccupied. Many of these unoccupied lots had been assigned previously by the *comisariado ejidal* to employees of the Ministry of the Agrarian Reform or sold to non-resident wealthy buyers. CORETT dismissed the agrarian delegate and appointed its own man on the post, refusing to give recognition to the sales and lot assignations that the *comisariado ejidal* had carried out in the upper section of the *colonia.* It began to sell to new buyers and intended to evict any previous settlers and claimants from the area.

There were also 250 settlers actually occupying lots in addition to a group of aspiring settlers who were doing weekend voluntary work under the *comisariado*, in hope of being assigned a lot. The group was backed by the *comisariado ejidal*, from whom they had obtained their lot rights. This heterogeneous group, organized itself in June 1976 into a settlers' association and elected its own officers, while the *ejidatarios* were negotiating their agreement with CORETT. When CORETT denied recognition to their claims and began to sell their lots, the *ejidatarios* called a meeting with the participation of the affected lot holders and the representative of the League of Agrarian Communities. They denounced CORETT's sales as a violation of the agreement and accused CORETT of corruption.

Although the PRI was instrumental in backing the activities of CORETT, a congress woman belonging to the party made her appearance in the upper section and raised hopes among the settlers of blocking the actions of CORETT in the upper section. With the help of a local leader, she started in March 1977 to raise money among the settlers to interpose an injunction against their eviction from the site, and began to take the settlers to mass demonstrations and meetings at various government offices. As the settlers from the upper section were being shuffled from one office to another without any visible results, she rapidly began to lose her credibility and soon vanished from the area when the DF government finally refused to bypass CORETT.

Spurred by the conflict in the upper section, the *ejidatario* opposition group created a special committee to pressure CORETT which also included some settlers of the upper section. New elections to the *comisariado ejidal* were held in early 1977 as the new López Portillo administration took office. The Tlalpan group finally came to power after years of struggle, and Oscar Delgado was reinstated in his post by the new administration as *delegado agrario*. With backing from the League of Agrarian Communities (CNC), he achieved CORETT's recognition for the 250 lots assigned to the employees of the Ministry of the Agrarian Reform. He, however, showed little concern for the other settlers which CORETT was threatening with eviction.

To strengthen its position, the new *comisariado ejidal* began an intensive new round of lot reassignments to the government bureaucracy in the upper section, while continuing to subdivide and sell lots irregularly. DAAC employees, PRI activists, and members of the Secret Service were assigned lots. The confrontation with CORETT, however, did not come to an end. The *ejidatarios* were once more protesting against CORETT in 1978, denouncing its speculation with lots and its sales activities in the Periferico section. At the end of that year the question of the final destiny of the compensation funds due to the *ejidatarios* was still unresolved, as neither CORETT nor FONAFE showed any readiness to give up control of these funds.

The *ejidatarios* had not only to contend with CORETT but were soon facing the threat of organized invasion by groups of settlers. A massive invasion was staged by around 300 *paracaidistas* on the unoccupied lots of the Periferico section in January 1977. The *comisariado ejidal* turned to the police for help, and an armed police detachment assisted by police patrols and helicopters evicted the invaders, in compliance with the policy of the new mayor, Carlos Hank Gonzalez (1976-82) not to tolerate land invasions in the DF anymore. The same determination was shown by mayor Hank Gonzaléz against 12 different invasion attempts carried out between February and October 1978 on the land that ISSSTE had acquired indirectly years ago from the *ejidatarios*.

Prices of land increased in the *colonia* with regularization, and with them the incentives to sell. While regularization undoubtedly contributed to that increase, much of it came about as a result of value increases in neighboring areas and the public and private investment and physical improvement activity that followed regularization. The regularization operation allowed local government to increase taxes, raising the cost of land ownership and the cost of living in the area. The tax increases and the new services fees became the cause of conflicts between residents and the city authorities.

The greater profit opportunities and the higher cost of living and of land ownership brought about some changes in the composition of the population and in the use of land. While many settlers stayed, some departed as they could not afford lot ownership under the new conditions, or because they found it attractive to sell and benefit from their equity. For other owners it became profitable or necessary to use their lots for rent-producing activities, changing the use of the lot and increasing its building density. As a result many rental *vecindades* made their appearance in the lower section of the *colonia* by 1980.

CASE STUDY 3: CIUDAD NETZAHUALCÓYOTL

The Ciudad Netzahualcóyotl study presented here is a synthesis of the case studies of various authors: Martín de la Rosa, María Teresa Guerrero, and Martha Schteingart.[39] It adds important additional insights into the formation of the irregular settlements in the MAMC. While the two other case studies, San Bernabé and Miguél Hidalgo, illustrate the evolution of the irregular settlements that were created under the *zona urbano ejidal* cover, the Ciudad Netzahualcóyotl study is representative of the other two types of irregular settlements: invasion and developer-sponsored settlements, which were discussed in earlier chapters.

There are other, no less important differences between this case and the previous ones. One of these differences is the location of Ciudad Netzahualcóyotl within another political jurisdiction, the State of Mexico, where the state and local governments were accessories to the creation of the irregular settlements. Another important difference is the central role played by developers and financial groups in promoting the irregular settlements, a role played by the *ejidatarios* and *ejido* authorities in the DF. Last, but not least, is the location of Ciudad Netzahualcóyotl on inhospitable land, totally unfit for urban development and the presence of conflicting property claims by private owners, *comuneros* and government on the land.

Historical Origins of the Area

Ciudad Netzahualcóyotl, the largest irregular settlement in the country is located on the desiccated lake bed of salty Texcoco Lake to the east of Mexico City, near the international airport. It occupies an area of 6220 hectares in the State of Mexico on its border with the Distrito Federal, 9% of the total extension of the MAMC. The desiccation of Lake Texcoco, which lies in the lowest area of the valley, already began in colonial times to protect Mexico City from periodic flooding by draining the Valley to the outer basin. Dessication was further accelerated since the Díaz years with the building of a series of major engineering works, intended to bring the flooding problem under control.

The tenure status of the land on which Ciudad Netzahualcóyotl has arisen is quite complex, reflecting many of the struggles for land that characterize Mexican history, as described in general terms in chapter IV. The first settlement in the area was the free Indian village of Chimalhuacán, which received its first *comunal* land grant from the Crown in 1570. The endowment was increased in 1709. Most of this land was not adequate for cultivation and was not used for agriculture.

The adoption of the Lerdo Law in 1856, which ended corporate property, brought about the subdivision of the *comunal* lands among the villagers, in what has been known since colonial times as *comun repartimiento*, by means of which each family in a commune would be assigned a individual lot, which according to colonial law remained the property of the community.[40] Some of the land belonging to the village was invaded by the neighboring *haciendas* but the Díaz government formally recognized in 1881 the validity of the land grants given to the village of Chimalhuacán in colonial times. When the Díaz government built the Xochiaca Dam in 1902 to continue the desiccation of the lake, new land gained from the lake became available. A new village, San Juán

Pantitlán, which depended administratively on the *ayuntamiento* of Chimalhuacan, was formed. Its residents were successful in adapting the land to agricultural activities. They draw a living from agriculture, fishing and insect hunting, selling their products in their own local market.

After the ousting of Díaz, the Madero government scrutinized in 1912 the legal status of the lands gained from the shrinking lake, and declared them to be national property under the jurisdiction of the Federal Government. A further relevant development came about when all natural inland lakes connected with flowing waters and their shores were declared to be national property in the Queretaro Constitution of 1917. The Chief of State Carranza ordered in 1917 a survey of the lake area to determine the boundaries of the national lands.

However, in what may seem to be a deviation from the former approach, the Carranza government published a resolution in 1919 declaring that the national lands in the Texcoco area were not being designated for any specific purpose such as forestation or colonization and could be sold at the price of $MP 60 an hectare. President Obregón again issued in 1922 a decree declaring all water bodies, streams, lakes, brooks, canals and ravines of the Valley of Mexico to be national property, but the land gained to the lake continued to be sold off by the government ostensibly for agricultural improvement to influential revolutionary politicians and military leaders as late as the early thirties. They were issued official property titles by government bodies such as the Ministry of Agriculture and Development. *Comunal* land belonging to the Chimalhuacán commune was also deeded in these transactions, giving rise to unending conflicts and claims by the *comuneros*, who for many years tried unsuccessfully to recover them in the courts.[41]

The decision to sell the desiccated lands was a most unfortunate one, to say the least, from an environmental standpoint, for these dried-out marshlands were totally inappropriate for agricultural or urban uses. The full meaning of this statement can be realized when the lunar landscape of Ciudad Netzahualcóyotl is contrasted with the remaining lake area still in government hands, which has become a world model for marshland restoration in recent years. This area has been successfully regenerated to its original natural role as wetlands, using recycled sewage water. It is now teeming with a rich aquatic life sustaining a chain of migratory and local birds.

When it became clear that these lands could not be improved for agricultural purposes, prices crashed in 1929 to one peso an hectare, and in 1931 to 25 cents an hectare. President Cárdenas tried to recapture—without much success—some of these lands for the public when it became clear that agricultural development would not take place on them.[42] He issued a resolution in 1936[43] declaring null and void the property titles that were issued on the basis of a previous presidential resolution (May 23, 1932). The owners resorted to a court injunction against the executive decree, which some of them won in court.

Thus at the end of the thirties, there were four types of tenure in the area, often overlapping and producing a state of tenure uncertainty: *comunal* lands with titles of *comun repartimiento* since 1856; *comunal* lands which had been taken away illegitimately; private lands sold or donated by the Federal Government; and national lands. Additional confusion into the tenure situation was introduced by the shifting position of the property lines facing the receding lake as a result of the advancing desiccation process.

Settlement of Ciudad Netzahualcóyotl

The unsuitability for urban development of the mud flat once covered by the salty Texcoco Lake could hardly have escaped the attention of even the least perceptive. The natural landscape of the barren mud flat lacked trees and vegetation as a result of the salty nature of the soil. In the dry season it became a dusty, desolate plane, in which the dust storms that contaminate Mexico City have their origin, while in the rainy season it turned into a muddy swamp. Its soft, exceptionally weak saturated clay soil, with its water table almost at the surface, made that soil inadequate as a foundation for buildings, and highly vulnerable to earthquake damage. Nevertheless on this desert, 1.5 million poor people, migrants, and former slum residents, were brought together to create a major settlement in just a few decades in what Francis Violich called "the most mechanized, economically and socially exploitative example of urbanization and urban design that one could find in Latin America."

When population pressures began to build up in Mexico City in the forties, private development companies bought up large tracts of land in Netzahualcóyotl for minimal prices.[44] Two types of owners, in addition to the *comuneros* and the federal government, characterized the area; one was comprised of owners that held land since the twenties, and the other—the larger group—were owners who perceived the great profit opportunities that were present in the area and began buying land or entering in partnerships with those holding any type of property claim there.

The developer companies acted with foresight and within a well coordinated plan from the business standpoint. They knew that the extreme environmental limitations and poor sub-soil conditions of the land made it unlikely that upper class subdivisions or heavy industry could be developed on the site.[45] On the other hand, they were fully aware that its relative proximity to downtown Mexico City would make this land attractive to the rapidly growing low income population of poor rural migrants and city slum dwellers, with their modest demands and low expectations, who were left without any housing option after the rent freeze.

The settlement process of CN went through four distinct stages: from 1945 to 1951 when the first settlements began by means of land invasions to circumvent a ban by the State of Mexico government against land sales and development; from 1951 to 1963 when the ban was lifted, and sale by the companies became the main route of settlement establishment; from 1963 to 1970 in which Netzahualcóyotl seceded from Chimalhuacán and the most intensive settlement activity took place; and, from 1970 to the present in which regularization was implemented.

The First Period (1945-51)

The first period transpired at a time when the effects of the rent freeze decree of the early nineteen forties were already being keenly felt by the rapidly growing low-income population in the form of a greatly diminished supply of new *vecindades*. The low income people, slum dwellers and migrants, began to search in the periphery for unserviced cheap land. Land invasion by settler groups was the typical method of settlement in the first period, as a means to force open the land for development to circumvent a ban on land sales which the State of Mexico kept in force between 1945 and 1951. There are conflicting versions of the events of those days on which settlers, *comuneros* and developers disagree. One such point of disagreement concerns the origin of the early land invasions in Netzahualcóyotl.[46]

According to the settlers version, the developers themselves were the promoters of the invasions to circumvent the sales ban that the state government had decreed between 1945 to 1951. Wishing to subdivide and sell the land, they forced urbanization by means of settler invasions and subsequent sales agreements, while at the same time protecting themselves through court injunctions against the settlers. Thus were created the first irregular *colonias* on the Texcoco lake bed: El Sol, Mexico, and Estado de Mexico, where the sales price was set between $3.0 MP/sq. m. ($0.62 US/sq. m.) and $5.0 MP/sq. m. ($1.03 US/sq. m.).

According to the developers' version, these invasions were prompted by large groups of settlers from the rural areas and later from the center city who invaded land that they, the developers, had acquired from *comuneros* who sold their lots of *comun repartimiento*, or from peasants that the federal government had endowed with land.[47] After the invasions the developers claim that they found themselves forced to negotiate with the settlers the sale of the land at minimal prices that would typically run around $MP 2.50/sq. m. The developers claim that with the exception of one *colonia* which they thus sold in 1947, sales really began in 1954.

The Second Period (1951-63)

The second period began after the sales ban was lifted in 1951 in the State of Mexico, and mayor Uruchurtu issued his ban in the DF. The State of Mexico border zone surrounding the northeastern periphery of the DF became the escape valve for the low income population. While in the DF, land developers played a minor role in the formation of the irregular settlements, developers became the catalysts of irregular settlement activity in this period in the State of Mexico. Developer-sponsored illegal subdivisions became the characteristic form of irregular settlement activity in the State of Mexico. Sales began in 1954. The developers organized business corporations that where capable of attracting substantial financing. Some held control of as many as 15 *colonias* in the area, in addition to their other holdings in the DF and in the rest of the country.[48]

The state government adopted a permissive, hands-off policy of unrestricted growth within the Estado de Mexico, turning the state periphery into a haven for the irregular developers. The new governor of the State of Mexico backed the irregular developers (1952-58), a policy which his successors continued during the sixties, allowing them to subdivide and sell unserviced lots in Netzahualcóyotl. The subdivision promoters moved in, and began to subdivide and sell the great extensions of barren land under their control. Being on the other side of the line of the DF, they were in a position to benefit from the business opportunity that the DF ban opened up for them.

Although in 1958 the State of Mexico adopted a new code that made the approval of any new subdivision conditional upon the introduction of basic services such as water, sewers, electricity and street paving by the subdividers, the development of unserviced *fraccionamientos* not only continued undisturbed but gained new momentum. Many developers were exempt from introducing services even after the new code was adopted. The companies were thus able to sell at low prices and still make an handsome profit. The lack of infrastructure investment combined with the minimal cost of the land explains the large profits that were made by the companies.

Colonia Azul is mentioned by the developers as a good example of the price levels in Netzahualcóyotl. Prices for a lot in 1954, started there at $MP 365.00 ($US 29), payable in 10 years. Five years later (1959) their sale price had risen to $MP 975.00 ($US 77) payable in 10 years.[49] 15 years later (1974) such lots were already selling at a price fluctuating between $MP 25,000 ($US 1,960), and $MP 40,000 ($US 3,137), roughly 100 times their original value.

A great number of unserviced settlements were developed and sold during this period, although the buyers actually settled on them in the sixties. CN grew by 1962 to a population of 62,000 people, just before its secession as an independent municipality and the beginning of its unprecedented growth.

The *comuneros* of Chimalhuacán contested in this period the property rights of the developers. They made various attempts to prove that the lands which the developers were selling were still *comunal* and national lands. They denied the legitimacy of the acquisition of *comunal* land of *repartimiento comun* by the subdividers, and argued that the *comunal* lands which were divided in the Juarez days in *repartimiento comun* were not marketable, and as such still belonged to the Community of Chimalhuacán. The struggle of the *comuneros* organizations, such as the independent Unión General de Obreros y Campesinos de México (UGOCM), to bring about restitution of the land to the *comuneros* did not bear any fruits. This organization nevertheless played an important role in later years in the massive protest movement against the development companies that led to the ouster of the companies, regularization, and minor compensation for the *comuneros*.

The Third Period (1963-73)

The third period, the boom period of irregular land subdivisions in Ciudad Netzahualcóyotl, began with the secession of the municipality on April 22, 1963. Netzahualcóyotl seceded from Chimalhuacan in 1963 to become an independent municipality, clearly for the purpose of initiating without interference its uncontrolled subdivision activity. The initiative for the secession of Netzahualcóyotl is credited in the official government version to the settlers, who supposedly were interested in obtaining for Netzahualcóyotl the status of a city to better take care of their collective needs as a independent municipality. The settler version differs, and credits the initiative to the developers as a means of securing control of the local government to facilitate their development activities. The developer version agrees that the initiative originated with them, although the reason given by them is that of freeing themselves from the hostile authorities of Chimalhuacán, who were unjustly imposing fines and violations at every step and turn blocking development of Netzahualcóyotl.

With the secession of Netzahualcóyotl in 1963, the developers began to operate on a grand scale. They held full control of the local government in the new municipality. Complicity between the State authorities and the companies reached a climax. They created, in that decade alone, 83 new *colonias* in Netzahualcóyotl. The incredible number of 150 thousand lots of 150 sq.m. each, were laid out in a mechanical fashion on a monotonous checkerboard, with a minimum level of services which included some electricity lines, a few stand-pipes for communal use and a few curbstones to delineate streets, and promises of future service introduction. The high rents in the DF and the comparably low prices of the lots brought in great masses of buyers. The rumor that the Metro would soon be coming to Netzahualcóyotl helped greatly to increase sales. It is estimated by some that the developers made in that period an estimated profit of 3000 million pesos, the equivalent of 240 million US dollars.[50]

Population began to grow exponentially in Netzahualcóyotl during the sixties. In the 1960-70 decade the population grew nine-fold to 580,436 people, (official census data, 1970). A 1970 population survey by the Ministry of Water Resources on the origin of the settlers of Netzahualcóyotl showed that 31% of the settlers migrated directly from the provinces to the area, 9.7% from the State of Mexico and the rest, 58.7%, had come from the DF. This last group, however, included many who were not DF natives but moved there earlier from the provinces, probably having lived in the center city *vecindades* before moving to Netzahualcóyotl.

Prices for consumer articles in Netzahualcóyotl were higher than in the rest of the city, and absorbed more than 50% of the family budget. The average individual and family monthly incomes in 1970 were $MP 955 ($US 76.4) and $MP 1228 ($US 98.2) respectively. Only 3.6% of the population had a monthly income equal or above $MP 3000 ($US 240), but most (62.9%) stood below $MP 1200 ($US 96). Families were typically large with an average of 6.2 members per family, higher than the national average. The percentage of unemployed among the adult population capable of work was higher than the national average, fluctuating between 45% and 56% in 1969-70.[51]

Although the developers argue that after 1958 they were complying with the subdivision code, the objective data do not bear out their contention. Large parts of the subdivisions lacked the most elementary utilities, as in the older days. In 1964 only 47% of the settlers had electricity, of which 31% was illegally connected. By 1968, 54% of the area still lacked sewers, and by 1971,

44% of the residents lacked water. With the exception of the main thoroughfares all the streets were unpaved. Later-day official government versions lay the blame on the developers, whom they accuse of systematically violating the development code with the backing of "immoral government officials who were in connivance with the developers."

As late as 1973 there were only 8 public and 30 private telephones and two post offices for nearly one million people. It was only until 1973 that the first gas station was opened in Netzahualcoyotl. Public transportation was one of the major problems in the area. Only three bus lines, run by a monopoly, served the 200,000 workers commuting daily to other parts of the city. The buses, in ruinous condition, would be packed to standing limit, and would even carry people on their roof tops. Bus lines to connect Netzahualcoyotl internally were non-existent.[52]

Drinking water was scarce, and people would rise at four or five o'clock in the morning to form long lines and wait for long hours at the public standpipes to obtain a bucket of water. During the dry months the women would rush from one *colonia* to another on the basis of unfounded rumors, looking for the rarely appearing water tankers to buy a twenty-liter container for five pesos. A foul smell permeated the area in these years due to the sewage that was discharged from the DF to the lake of Texcoco. The insufficient number of garbage collection trucks left much of the garbage uncollected.[53]

During these years Netzahualcóyotl reached one of the highest mortality rates in the country, due mainly to intestinal and respiratory ailments. Child mortality was particularly critical. The child mortality rate in Netzahualcóyotl was more than double that of the State of Mexico (227.3 vs. 105.5). Half of all the registered deaths were in the category of children of less than four years old and one third of less than one year old.

Around 1969, in a watershed year of Mexican politics, the residents created a defense organization against the development companies, the "Movimiento Restaurador de Colonos" (MRC), whose struggle against the companies was based on the claim that their property rights on the land were illegitimate and that they had not complied with their promises and the legal requirements of services. They challenged the legitimacy of the developers' property rights on the land, which they claimed belonged to the *comuneros* of Chimalhuacán and the state. They organized mass demonstrations against the developers and declared a payments strike, which spread to other irregular settlements across the city. They demanded expropriation of the land by the government and jail for the developers for fraud in selling land that did not belong to them, and for not complying with their sale promises of services. Their movement finally led to government takeover of the companies and to implementation of a regularization program in Netzahualcóyotl in 1973.

The Fourth Period: Regularization (1973-)

The settlers mass movement is linked to a sharp reversal in government policy on many issues, including company-sponsored development. The MCR movement grew with the new political winds that started blowing as a result of the so-called *apertura democratica* "democratic opening" policy of the new Echeverria regime.[54] The alliance with the development companies was given up by the government for a policy of state takeover and populism that was very much within the political philosophy of the new administration of President Echeverria (1970-1976). The MCR movement gained much strength when it urged the settlers in 1970 to suspend payments to the companies and deposit these instead in Nacional Financiera so that the state government could invest these funds in infrastructure.

By 1971 MCR claimed a membership of 70,000 settlers organized in 48 subcommittees. The government rapidly stepped in and set in motion a strategy of cooptation of the MCR movement. "Promises, economic rewards, and public posts for the leaders of the movement," soon divided the movement and brought it under full control of the government. On the basis of the claims of the *comuneros* of Chimalhuacan to the land, MCR was incorporated in August 1971 into the Confederación Nacional Campesina (CNC), one of the three sectors of the government party representing peasant interests.[55]

The government came forward with a proposal to create a trust fund which would take over the companies and handle the regularization of the property rights and the introduction of services in Ciudad Netzahualcóyotl. In March 1973 an agreement was signed between the state government, Nacional Financiera, the development companies of 43 of the 83 existing *colonias*, and representatives of the settlers and the *comuneros* of Chimalhuacán, to create a trusteeship, the Fideicomiso Ciudad Netzahualcóyotl.

According to the UGOCM that agreement was reached against the will of the majority of settlers and *comuneros* who had come to the conclusion that the solution offered was more in the interest of the government and the development companies than in their own interest. The government, according to this version, nevertheless handled the situation successfully by winning over the representatives of the settlers and the *comuneros* for the agreement.

The agreement was originally established under the presumption that the property rights of the development companies would be scrutinized. A legal inquiry might have led to a recognition of the claims of the *comuneros* and the implementation of a CORET regularization program, requiring expropriation of the land as *comunal* property and resale to the settlers. The government, however, did not pursue such a course of action, which implied voiding the legitimacy of the property titles that previous administrations had issued to the private owners. This would have incriminated the earlier administrations as perpetrators of fraud. Such a course of action against previous administrations would have been most unlikely. A key political figure of the moment, Augusto Gómez Villanueva, head of the Departamento Agrario, declared: "The truth is that the developers brought forward property titles that were issued in the past by the government of the Republic; titles that General Cárdenas tried to nullify, and on which the Supreme Court passed judgement ."[56]

Unlike what the government decided to do in the ZUE settlements, it opted for totally bypassing the question of the legitimacy of the property rights of the developers and simply recognized and took over the sales and obligations that the companies had transacted. A contract was signed whereby the developers transferred to the Fideicomiso Netzahualcóyotl all their assets which included the property rights of the developments, some tracts of unoccupied land, and an outstanding debt by the settlers of $MP 623,607,249 ($US 49,888,580).[57] The settlers were granted a 15% discount on their outstanding debt, which was taken over by the trust fund. The development companies were granted 40% of the funds after setting aside $MP 50,000,000 ($US 4,000,000) to compensate the *comuneros*. The remaining 60% of the debt was earmarked for community services.[58]

A technical committee was established with representatives of the *comuneros* and the settlers to administer the investments. Payments by the settlers were once again renewed. The trust fund began by investing the funds due to the *comunero* group—which had meanwhile grown to 3075 claimants—in building a clothing factory which could offer employment to 100 workers. The funds due to the community were used to build a central plaza in Ciudad Netzahualcóyotl.

The main purpose, however, of the Fideicomiso was that of producing property titles for the settlers in the 43 *colonias* under its control. The Fideicomiso encountered such problems as conflicting claims of title by different buyers on the same lot. Lots had been sold more than once by the companies to different people; it also inherited hundreds of pending legal civil suits against settlers. A year after the creation of the Fideicomiso the first 500 property titles issued by the Fideicomiso were handed over to the settlers by President Echeverria on May 10, 1974.

New opportunities for profit were opened for the leaders of the movement, who were given free hands. They organized invasions of unoccupied lots, which often had owners and obtained payments varying between $MP 1000 ($US 80) and $MP 4000 ($US 320). More than twenty thousand vacant lots in Netzahualcóyotl were thus occupied through their instigation after the creation of the Fideicomiso until 1974, under the claim that the government ordered them to settle Netzahualcóyotl.[59] The Fideicomiso confronted these situations by such solutions as division of the plot between occupant and claimant, or by requiring the occupier to compensate the original absent owner.[60]

The federal and state governments undertook a major investment program financed by international loans to introduce services in this vast wasteland, and to bring an end to its periodic flooding. This last objective was finally achieved in 1972 by building a completely new, complex sewer system. A major investment program of 1,200 million pesos ($US 96,000,000) on infrastructure was implemented in 1973. In a dismally poor municipality where the total annual budget of the local government was then a merely $MP10 million ($US 800,000, an average of 80 US cents/resident), such an investment was bound to produce an important impact.[61] An average debt of 10 thousand pesos per lot ($US 800) was assessed on each owner for the recovery of the government investment. Although the settlers felt that great progress was being made, they objected to paying for what they considered was the developers' obligation. They also protested the high prices of the infrastructure investments which they considered to be blown-up.

Population growth in Ciudad Netzahualcóyotl took on new momentum and reached 1,341,230 people by 1980. However, according to the 1990 census it decreased somewhat during the last decade to 1,256,115, due probably to its rapid expansion on nearby Chimalhuacan. Netzahualcóyotl, with its 110 colonias, is still to this day the most eloquent example to be found anywhere of the lasting effects of unplanned development. Social and sanitary problems are still acute, and police corruption and repression is cited in different surveys of the area as critical. More than half of the streets are still unpaved. At present, as a result of the lack of sanitary facilities in many dwellings, and the presence of 900,000 dogs and animals, epidemics are spread by dust storms carrying many tons of human and animal feces in the winter season.

Netzahualcóyotl is now the greatest single irregular settlement in the country, and very probably of Latin America and the Third World as a whole, having become the very epitome of the problems that characterize these settlements.[62] Neither the heavy government investments nor title regularization have been capable of remedying the original oversights such as its location on a dried lake bed, providing solid arguments against the policies and arrangements that can be held responsible for this type of growth.

CONCLUSIONS FROM THE CASE STUDIES

The three case studies provide a first-hand account of the deleterious effects that the lack of full property rights produced on the urbanization process in the MAMC periphery. The absence of this vital element was instrumental in one way or another into turning the urbanization process into a lifetime ordeal for *ejidatarios* and settlers. The enormous social costs linked to irregular urbanization which clearly outweigh any benefits one might find, are in great measure traceable to the problems of undefined property rights.

The conclusions that one is bound to draw from these case studies clearly points to institutional failure. The Mexican political system has failed to integrate the *ejidos* in the urban peripheries into the process of urban expansion. The *ejidos* have increasingly become an obstacle to urban growth and to public processes like planning since 1940. The pervasive illegality in the urbanization of *ejido* land and the occupation of much of that land by irregular settlements at the urban periphery continues to be the clearest symptom that the *ejido* legislation, whatever its merits for the rural areas, causes inefficiencies and inequities on the urban peripheries. The original restrictions of the *ejido* property rights in the rural environment lost any trace of functionality in the urban setting and became instead a formidable obstacle in the unavoidable process of urbanization.

The *ejidos* themselves have been the first casualties of this situation. Instead of being beneficiaries of urban growth, they have become its victims: a tempting resource to be minced by various national political interests who manipulated imprecise laws and thereby controlled the land use conversion decisions. One reason for this failure is that the Agrarian Reform, as originally conceived, did not provide the foundation for flexible and constructive adjustments which would integrate *ejido* land into the urban expansion process. Planned development or the achievement of other social objectives in the urban setting were clearly made impossible. This has become the source of continuous social conflict and physical problems in the expanding cities. Evidence of this is plentiful from the communities described in the case studies for over fifty years, as well as from the continuous belligerent response by the *ejidatarios* to present day expropriation initiatives reported almost daily in the Mexican press.

Naturally, the drafters of the original *ejido* legislation could not have envisioned the enormous urban growth pressures that have forced city expansion and engulfed many *ejidos*. Neither market processes, planning attempts, nor different government policies have been able to overcome the fundamentally flawed characterization of *ejido* property rights. Indeed, the very weakness of the *ejidatarios* with respect to their status as property owners within the land use conversion process provokes negative behaviors both of the statist elite, who stand to profit greatly, and the landed *ejidatarios*, who stand most often to lose. These behaviors contribute to environmental damage and to the breakdown of most steps toward responsible urban planning.

---

1 (Durand, 1983).

2 (María Luisa Medina Lúgo and Fernando Bejarano González, 1981).

3 (María Teresa Guerrero *et al.*, 1974; Martín de la Rosa, 1974; Martha Schteingart, 1981).

4 There are significant archaeological remains in the area.

5 The *hacienda* produced maize, beans, barley and fruits for the Mexico City market on its seasonal land.

6 The *hacienda* employed an estimated 100 day-laborers during the four-months growing season. In addition to the seasonal workers from the village, the *hacienda* had a small permanent labor force which cared for the orchards, cattle, forests and the *hacienda* house.

7 In the small town of Contreras, 2 km. from the village, there has existed textile production since colonial days. The textile factory "La Magdalena," existed from the first half of the XIX century until recent times when the factory closed. Other textile and paper mills were established in Contreras and in the nearby town of Tizapan around 1880.

8 A census undertaken by the agrarian authorities on the 19th of December, 1922 to prepare the land endowment program showed that the active labor force of the village totalled 179 individuals of which 126 were salaried employees. In addition to 53 peasants there were 71 agricultural day laborers, 19 textile workers, 1 bricklayer and 35 female employees who worked as domestic maids in Mexico City. In the 1923 census there is additional reference to 13 craftsmen and 1 merchant.

9 The appearance of new aspiring *ejidatarios,* may be due in part to the coming of age of 45 minors, who in 1924 had not reached the necessary minimal age, while the rest may have been encouraged to become farmers by the policies of the Cárdenas era.

10 New employment opportunities in the building sector expanded greatly in the southern tip of the city: The "Jardines del Pedregal" residential section started in 1949, the UNAM campus was begun in 1951, " Unidad Independencia" at the end of the 50's, the World Olympics and the World Soccer installations in the sixties.

11 (DO, March 25, 1954).

12 (DO, August 27, 1975).

13 The Durand case study stops at this particular point. The later developments have been gathered directly from the Diario Oficial.

[14] (DO, March 17, 1981).

[15] (DO, July 20, 1989).

[16] (LFRA, art. 125)

[17] (Lugo Medina & Bejarano González, 1981).

[18] A trolley car line linked San Agustin de las Cuevas to Mexico City and a railroad line linked Tlalpan with Tacubaya, Mixcoac, San Angel, Mixcoac, Tacubaya and Mexico City.

[19] Land belonging to the peasants of the Rancho del Teochihuitl was donated by the town government of San Agustín de las Cuevas to the parish church, which leased this land to private persons. The peasants began a long struggle to recover these lands. They suffered a serious setback in the days of the Juarez government (1861) when these lands were publicly sold as a result of the Lerdo laws, to a private business firm.

[20] The peasants of Rancho del Teochihuitl accused a neighboring factory in 1869 of taking away land and waters from the "Fuentes Brotantes" to their rancho. They negotiated a partial settlement in 1884 in the days of the Díaz government, by which the factory allowed some of the water to flow for the irrigation of their land.

[21] (DO, January 28, 1930).

[22] (DO, January 28; May 4, 1938).

[23] In the period between 1949 and 1974, 1957 hectares of *ejido* land were expropriated in the combined Tlalpan Contreras area for public interest purposes and 1083 hectares were legally bartered for upper class developments .

[24] (DO, August 2, 1944).

[25] (DO, September 25, 1946).

[26] (DO, June 2, 1954).

[27] (DO, May 19, 1959).

[28] The first section of Jardines del Pedregal had originally been established through a barter operation on 125 hectares of land of the neihgboring *ejido* San Jerónimo Aculco in the early fifties (DO, January 18,1950; May 14, 1951).

[29] The same year in which the barter operation was approved, 1959, President López Matéos (1958-64) created the National Fund for *Ejido* Development (Fondo Nacional de Fomento Ejidal-FONAFE), a national trust fund to plan and administer the investment of the common funds of the *ejidos*.

[30] Although land barter at least allowed some of the *ejidatarios* who agreed to resettle to maintain their agricultural occupations, it was outlawed in 1970 by President Echeverria (1970-76), seemingly to make government the main beneficiary of *ejido* expropriation.

[31] (DO, October 29, 1951).

[32] (DO, May 19, 1959).

[33] The quotas varied between $MP10 and $MP 20 to be used in the further subdivision of the area.

[34] The authors list a number of additional privileged bureaucratic recipients which includes agents of the secret police, agents of the ruling party, and employees of the national railroads.

[35] The *ejidatarios* accuse these promoters of having defrauded them in the sale operation, without specifying in what way was this accomplished.

36 Settlers believe that the deposed comisariado's withdrawal of recognition of the lots assigned to the DAAC bureaucracy triggered these actions.

37 The *comisariado* had the support of the agrarian bureaucrats to whom it had assigned lots in the settlement and also of the aspiring settlers who were participating in the weekend work assignments in the upper part of the *colonia* in hope of being assigned a lot.

38 Libro de Valores Inmobiliarios del Valle de Mexico, ed. por el Instituto Mexicano de Avalúos, A..C. Mexico 1975.

39 (Martín de la Rosa, 1974; María Teresa Guerrero, 1974; Martha Schteingart, 1981).

40 See chapter IV for a more detailed discussion of *comun repartimiento.*

41 (Schteingart, 1981).

42 (Guerrero *et al.*, 1974).

43 (DO, March 28, 1936).

44 (Guerrero *et al.*, 1974).

45 (Gilbert & Ward, 1985:p. 87).

46 (Schteingart, 1981: p. 107-109).

47 Ibid

48 Ibid.

49 With a down payment of $MP 15.00 ($US 1.17) and monthly payments of $MP 8.00 ($US 0.63).

50 (De la Rosa, 1974).

51 Ibid.

52 Ibid.

53 Ibid.

54 Ibid.

55 The first internal split in the MCR movement occurred when the leader actually joined the government party and was soon cooperating with the municipal president whom he previously had attacked. A second more serious rift occurred at the end of 1972, when from seventeen members of the executive council, fourteen ended up being coopted by the government, which needed them to push through its solution to the problem.

56 Ibid.

57 (DO, May 15, 1973).

58 President Echeverria declared in his visit (May 10, 1973), just before the official publication of the Fideicomiso agreement: "The trust fund made it possible that the money which the settler earns with such great effort, should not go in its entirety for the developer. Unfortunately, we have to hand him over part of it, because that is what the laws of Mexico say; we can not from the government violate the law without falling into anarchy." The governor Carlos Hank González declared when referring to the companies: "We should not even give them one nickel, but that is legally difficult." (De la Rosa, 1974; p. 23).

59 "So powerful were local community factions that the Trust was forced to give local 'committees' the right to allocate remaining parcels. In El Sol the plots were distributed by four committees and most residents acquired plots by paying a set contribution of 2000-3000 pesos to the leaders. A few others simply invaded. In many cases those who had

bought in the 1960s arrived in the settlement in the early seventies to claim their plots only to find them already occupied." (Gilbert & Ward, 1985:p. 282-283)

60 The MCR leaders rapidly began to grow rich with cars and armed body guards. They became involved in selling "protection" to the owners of the furniture credit stores in the area, who were threatened with a general payments strike. Their most violent mass action however, took place in the beginning of 1974 when they inflamed the people against the bus companies which had increased their fares before the new rates had been officially authorized. That action climaxed in mob killings of various bus drivers on January 5, 1974 (De la Rosa, 1974).

61 400 million pesos were spent on the sewer system, 120 million on electricity, 44 million on drinking water, and 100 million on other works. One tenth of the street network stretching over 1100 km was paved.

62 Report by the National Anti-Rabies Institute published in Excelsior, January 29, 1989.

Chapter VI

POLICY CHANGE PERSPECTIVES

The realization that policy changes are required to allow the planning system to manage the process of urban expansion and counter the negative effects of *ejido* informal urbanization has already prompted important discussions in political and academic circles and given rise to policy proposals that require careful consideration. These discussions and proposals will be surveyed in this chapter, while it is argued that they have left the basic question of the *ejido's* blocking effect unresolved. Such a survey should provide the necessary backdrop for the policy proposal to be presented in the next chapter. Before commenting on these discussions it is necessary, however, to refer briefly to the planning framework within which any policy proposal must be set and take cognizance of the possibilities and limitations of the legal infrastructure of planning in force now.

THE PLANNING FRAMEWORK

The present constitutional basis for urban planning at a national scale in Mexico was laid in the middle seventies. A proposal to amend articles 27, 73, and 115 was sent to Congress on November 12, 1975 at the end of the Echeverria administration, making it a public responsibility to order, plan, and regulate the human settlements and population centers. The need to prepare land reserves in anticipation of urban expansion, instead of allowing and fostering irregular settlement and chaotic growth, was given official recognition in the Mexican Constitution.[1] The constitutional amendments served as the basis to the Human Settlements Law, which was presented to Congress a month later on December 15, 1975 and which was approved on May 20, 1976.[2]

The amendment to article 27 (paragraph III) of the Constitution, which is highlighted in bold typeface, reads as follows:

> The Nation shall at all times have the right to impose on private property such modes (modalidades) as the public interest may demand, as well as the right to regulate to the benefit of society the utilization of the natural resources which are susceptible of appro-

> priation, in order to make an equitable distribution of the public wealth, take care of their conservation, achieve the country's balanced growth and the improvement of the living conditions of the rural and urban population. **Consequently, the necessary measures shall be dictated to order the human settlements and establish adequate provisions, public and private uses and reserves of lands, waters, and forests, in order to carry out public works, and plan and regulate the foundation, conservation, improvement and growth of the population centers;**[3] to divide the latifundia; **to determine through its bylaws the organization and collective exploitation of the *ejidos* and communities;** to develop small landed holdings; to create new centers of rural population with the lands that are indispensable to them; to encourage agriculture and to prevent the destruction of natural resources and the damages that property may suffer to the detriment of society. Villages, hamlets situated on private property, and communities which lack lands and water, or do not have them in sufficient quantities for the needs of their inhabitants, shall have the right to be provided with them from adjoining properties, always having due regard for small landed holdings. Therefore, all grants of land made up to the present time under the decree of January 6, 1915, are confirmed. Private property necessary for the achievement of the above-expressed objectives shall be considered as taken for public utility.

The introduction to the amendment states the various objectives of the law, one of which is the achievement of a more equitable and harmonic pattern of regional development, which should allow the evolution from a macrocefalic system of cities to a less concentrated pattern of urban centers leading to greater productivity, economic growth and social justice. Explicit reference is made in the introduction to the fact that urban expansion is not keeping pace with population growth creating a housing shortage which mostly affects the popular sector. Uncontrolled urbanization in the urban peripheries occurrs often on *ejido* and *comunal* as as the formal process of land conversion breaks down. The pressure of all those who have a need of a roof and public services has led to the formation of irregular settlements not subject to any order which hinders the public authorities' responsibility of taking care of their needs in their locations.

The introduction to the amendment explains the need for institutionalizing an urban planning process at the 3 levels of government, federal, state, and municipal, each of which should intervene in the process according to the powers granted to them by the Constitution. An hierarchical three-tiered system of planning to operate at the national, state, and municipal levels was prescribed in the Human Settlements Law that followed. Each of these ranks of government was required to produce a *Plan de Desarrollo Urbano* (Urban Development

Plan). Three types of plans were mandated: a National Urban Development Plan, State Urban Development plans, and Municipal Development Plans. The lowest level municipal plans were made subordinate to the state plans and these to the national plan. The plans were charged with the responsibility of anticipating the demographic, economic, and physical growth of settlements and providing guidelines on how to respond to changes. Where cities expanded into more than one state it required that interjurisdictional planning commissions be established to coordinate the plans of the different entities and levels of goverment.

The urban development plans were to be followed by *Declaratorias de Reservas Territoriales* (Land Reserve Declarations), to be issued periodically for each city to prepare sufficient land for its growth and expansion needs. The land reserve declarations geographically defined the areas to be used for expansion and those to be left in agricultural use or as ecological reserves. The plans as well as the land reserve declarations were to be mandatory for five year periods, at the end of which they were to be updated.

While the amendment established a federal responsibility for ordering and planning the human settlements and population centers, it did not grant the state any new powers to abolish private property in carrying out the plans, but only applied the general principle which was proclaimed in 1917 that the Nation had a right to impose limitations (*modalidades*) on private property when planning the human settlements. The principle shared by the 1856 and 1917 Constitutions that private property could not be confiscated and could only be expropriated by means of compensation on grounds of public interest causes remained in force.

Although the power to impose restrictions on private property on grounds of public interest could lend itself to different interpretations and could lead to arbitrary abuse of power by the state in creating and applying the urban plans, safeguards to prevent or at least restrain any capricious application of the plans have a precedent in the Mexican judicial system. In a decision given on November 24, 1971 at the beginning of the Echeverria administration, the Mexican Supreme Court after recognizing that the state was acting legitimately within its constitutional powers in limiting and controling land use through its master plans, stated that in every specific case where planning limitations were applied, government was under obligation of "justifying the need for the imposed limitations which should neither be capricious nor arbitrary."[4]

Nevertheless, the legislative initiative stirred a passionate political controversy at a time of great tensions in Mexican political life. An important group of Mexican industrialists and businessmen came forward against the initiative and expressed distrust to its intentions and objectives. Various authors have sought to uncover the reasons for such strong antagonism to a set of principles that in their esence have since a long time been part and parcel of the body of legislation of every western democracy. One author suggested that much of that antagonism of a part of the private sector can only be understood within the wider context of its response to the total record of the policies of the Echeverria regime. According to this view opposition to the initiative was not due to its contents but to its sponsor and the statist ideology that he tried to advance from his position of power. The opponents saw in the initiative "one more manifestation of a dangerous advancement of the state at the expense of free private initiative."[5]

Azuela de la Cueva[6] has suggested that the inclusion of a clause within the amendment giving the state the power to determine the "organization and collective exploitation of the *ejidos* and communities," and the populist rethoric of the introduction of the Human Settlements Law condeming land speculation and the accumulation of urban property in a few hands, *latifundismo urbano*, opened the door for the suspicion that the government had in mind a land reform program within the human settlements akin to the agrarian reform which distributed the *haciendas*. The same author also points to a clause seemingly overlooked by the opponents introduced within the Law of Urban Development of the DF, which granted the DF government the power of setting the maximum sales prices of the lots when authorizing subdivisions.[7]

Prompted by the federal government all state legislatures, with the sole exception of the state of Nuevo León, approved within one year their own state laws of urban development, laying the necessary legal foundations for the operation of a new ministry of urban planning, the Ministry of Human Settlements and Public Works (SAHOP) which was charged with the responsibility of implementing the law nationally. Planning of human settlements and public works was put under the same roof to achieve greater coordination between urban planning and infrastructure investment. The new ministry was given substantial financial and human resources to implement the Human Settlements Law and to administer the land reserves program.

Puting aside any speculations on the question whether there were hidden political agendas involved in the conflict and moving beyond contextual political considerations that may have prompted the opposition, the amendment can be seen today sixteen years after its approval in a different light. An objective analysis of the amendment and the HSL itself based on its content, reveals that the amendment in the final balance is formulated appropriately in a broad manner that leaves ample room for interpretation. Much as one can expect differences of opinion as to the means to achieve its stated goals, the principles underlying the amendment and the Human Settlements law are hardly contestable. A commitment of this nature seems entirely appropriate in a situation of rapid urbanization, even in a poor country. By establishing the need to anticipate urban expansion and prepare the necessary land resources for its occurrence, it offers a rational alternative to random and irregular land development. As a constitutional mandate, it sets general standards for the planning of human settlements in Mexico.

The legislation is more than anything else relevant to the public sector itself, whose programs and actions in the realm of public investments and use of land are often inconsistent and discordant in space and time. The different public agencies, levels of governments and administrations in succession, often act in disonance and require instruments of coordination such as provided by the urban development plans. The planning process prescribed by the amendment and the Human Settlements Law is required to bring in a sense of continuity from one administration to the other and a sense of harmony between the three levels of government, federal, state and municipality as between the different agencies within each of these levels of government. Such mandatory coordination through territorial planning is particularly pertinent when one considers the expansion needs of the cities and the fact that at least half of the land in the urban peripheries is *ejido* land which can only be put to urban use through the expropriatory powers of the President himself.

There are still missing, however, some crucial links which would really allow to implement the land reserves prescribed in the Human Settlements Law. Many of the reserve's important characteristics, such as where the land should come from, who should own the reserve, and how often it should be updated, were left open. No explicit roles for *ejido* lands in setting up the urban reserves were mentioned, despite the dominance of *ejido* land at the rural periphery of cities throughout Mexico, and in Mexico City in particular.

Additional legislative developments relevant to human settlements planning, took place in the years 1983 and 1984 during the presidency of Miguel de la Madrid. To carry out his municipal reform program President de la Madrid amended article 115 of the Constitution and transfered attributions held by the state governments to the municipalities. The amendment[8] granted powers to the municipalities to formulate, approve and implement their urban development plans and zoning plans; to create and administer their land reserves; to control and oversee land use within their territorial jurisdictions; intervene in the regularization of tenure of urban land; grant building permits and licenses and participate in the creation of ecological reserves. Fraction IV of the amendment also declared the local governments, the *ayuntamientos*, to be the recipients of the property taxes accruing from development, subdivision, consolidation, transfer and improvement of land as well as those based on the increases in value of real estate properties. The Human Settlements Law were modified to conform with the previously enumerated changes of article 115 that made the municipalities the main agent in planning and managing their urban development process. All the state legislatures introduced these changes into their own constitutions to be consistent with the federal Constitution.

The states on their part dealt with the question of how to harmonize the municipal plans with the state plans in keeping with the fundamental principle of the 1976 human settlements legislation. In the specific case of the State of Mexico where this question has the greatest significance as a result of the MAMC expansion on its municipalities the SM government instituted a series of measures to achieve this objective. The validity of the municipal plans was made contingent upon their approval by the state legislature and the issuance of a statement of congruity with the state's Urban development Plan by the state executive branch. The state executive branch was also assigned the power to single out specific urban areas and designate them as "Strategic Population Centers" in the State Urban Development Plan. The executive branch was made responsible for producing the local plans for these centers with the concurrence of their local governments or *ayuntamientos*. In addition to the approval of the state governor and the *ayuntamientos*, final approval by the state legislature to these local development plans was made mandatory. Azuela de la Cueva views these measures critically as an intrussion on the rights granted the municipalities by the amendment to article 115, but abstains from suggesting alternatives to achieve the vital objective of consistency between the two planning levels.

President de la Madrid also introduced some amendments to the Agrarian Reform Law related to urban expansion. Although the legal basis for the exercise of eminent domain to expropriate *ejido* land for specific public works needs was established long before the Human Settlements Law, the *ejidos* could not be expropriated for global urban expansion needs such as the urban reserves—as seemingly implied in the Human Settlements Law—until new amendments to the Agrarian Reform Law were adopted. This was done in 1984, finally settling an old discussion on the legitimacy of such an action, questioned by intellectuals on the left side of the political spectrum. The constitutional amendment to article 27 became the legal foundation to the additions to the 1984

Agrarian Reform Law in this respect. The Human Settlements Law was also modified to grant preferential rights to the municipal and state governments to acquire private land parcels included within the reserve declarations when these were being put to sale or where their property rights were being transfered by any other legal means.

While the 1976 constitutional legislation mandating settlement planning and establishment of land reserves should be understood as a long-term goal that can not be judged on the basis of short-term results, it is nevertheless true that the legislation it elicited did not further its objectives. Neither the Human Settlements Law nor the additions to the Agrarian Reform Law have been successful in incorporating *ejido* land into the planning effort as intended. These legislative actions instead induced a frenzy of illegal *ejido* sales and an increase of irregular settlement activity. Land use controls and the creation of urban land reserves for planned growth have not materialized. The inequities of *ejido* land expropriations have stood in the way of their implementation.

Although the failure of urban planning to influence urban expansion has many dimensions as indicated in the first chapter, it has become clear that the inequities of *ejido* land expropriation are centrally important to this outcome. The growing awareness of this conclusion prompted many planners and officials to put the question on the public agenda for discussion for the first time in 1981 at a national conference in Tlaxcala. A review of the public policy proposals presented there is necessary to set the stage for a discussion of a different approach to the problem which is advocated by the author in the last chapter of this book.

THE TLAXCALA CONFERENCE

The Tlaxcala Conference organized by SIAP-SMP took place in the city of Tlaxcala in 1981.[9] The conference can be considered an important landmark because it put the question of the erratic expansion of the cities on the suburban *ejidos* in a public political forum for the first time, and initiated the search for innovative approaches to the question of *ejido* conversion in the urban growth regions. The Tlaxcala conference was prompted by the challenge made during the presidential campaign of President De la Madrid (1982-88) to discuss new approaches and policies to the land problem in the urban peripheries. As much as one may differ with these approaches and policies, the conference should be credited with putting the question of the suburban *ejidos* on the political agenda.

Most papers presented seem to share a general commitment to the basic principles of planned urban growth expressed in the 1976 amendment to article 27. While concerned with the housing needs of low income people, many of these papers were critical of the government's "hands-off" policy, which allowed illegal settlement to take place mostly on *ejido* land in the periphery of many cities in the country. The conferees disapproved of regularization as the way to deal with the problems of *ejido* land conversion to urban use. Standing by idly and stepping in only after settlement has already taken place on these lands would lead, one participant argued, to "more and greater Netzahualcoyotls in the future." The conference generally seemed to adopt the position that it is government's responsibility to anticipate the establishment of the settlements. Planning, not ex-post regularization, was seen to be the road to be followed by the major cities, especially those undergoing expansion.

Arturo Mier y Terán, one of the participants, dramatized the outcomes of unplanned urban growth and regularization when he pointed out that, as the conference was taking place, a new Netzahualcóyotl was being created in the Valle de Ayotla, to the east of Mexico City. As many as 60,000 new irregular settlers a month were establishing themselves haphazardly in 20 different settlements on an estimated area of 2,500 hectares of *ejido* land.[10]

Conferees who rejected regularization in favor of planned development focussed on how to institute a more adequate urban conversion system of *ejido* land. Two basic approaches could be distinguished in these proposals: one approach was in favor of indefinitely maintaining the *ejido* custody under much stricter controls, allowing only development that could be accommodated within the narrow limits of traditional *ejido* legislation; the other approach saw the *ejido* regime as an insurmountable obstacle to planned urbanization and advocated the transfer of the *ejido* property rights to the government in the areas where urbanization is anticipated. None, however, suggested a conversion system leading to privatization of the *ejidos* in the suburbs.

Joaquín Alvarez Vázquez supported the first approach. He was in favor of maintaining *ejido* custody and property status of the land in the urbanizing periphery. His proposal argued for a unified law of territorial property, encompassing all types of ownership, to bring an end to the conflicts between the agrarian and the urban legislations. It is, however, difficult to see how he intended to accomplish this by tightening even more *ejido* non-marketability. He shared the unwarranted assumption of some circles that expropriation of *ejido* land for general purposes of urban development was contrary to the spirit of article 27 of the 1917 Constitution. The only valid public interest purpose that could justify *ejido* land expropriation, according to his view, was the one then already recognized in the Agrarian Law—public works. He contended that private uses—residential or industrial—should not be recognized as legitimate reasons to expropriate *ejido* land.[11]

The only place in the *ejido* where Alvarez Vázquez allowed room for some form of urban development and limited private property was in the *zona urbano ejidal*. However, even there he expressed his opposition to granting unrestricted private property titles to the ZUE residents as established in the agrarian legislation: families should instead only be granted *patrimonio familiar* titles,[12] to keep the land outside of the market. To minimize speculation, cooperative housing groups should be given priority in entering the *zona urbano ejidal* areas. Within the ZUE, according to this proposal, only *ejido* enterprises should legally be allowed to run the various community services, such as, gas stations, markets, cinemas, amusement parks, cemeteries etc.; participation of private interests in these enterprises should not be allowed to exceed thirty percent of their total capital value.

One could hardly find in the Alvarez Vázquez proposal any new elements that could contribute to reducing illegal and unplanned growth on *ejido* land besides stricter controls. Alvarez Vázquez clearly showed little sensitivity to the growth needs of the cities surrounded by land that primarily belongs to *ejidos*. Sticking to a rigid interpretation of *ejido* legislation as it is applied in the static conditions of the *ejidos* in the rural areas, he refused to recognize the radically different conditions in the urban peripheries and the enormous market pressures that act there. While one can hardly doubt that *ejido* legislation at that time only allowed *ejido* expropriation for public works, it is a far cry from arguing that recognizing other purposes in the law such as general urban expansion needs was antithetical to the spirit of the 1917 Constitution. The amendments introduced to the Agrarian Code after the conference in 1984 settled this discussion and gave legitimacy to the expropriation of *ejido* land for general urban growth purposes and establishment of urban land reserves.

Francisco Covarrubias Gaytán, former undersecretary of Urban Development at SEDUE, expressed in one of his two conference proposals some concerns similar to the previous author, although showing a far greater sensitivity to urban growth needs. He considered the possibility of relaxing the present ZUE restrictions and legally opening the *zona urbano ejidal* to certain groups of non-*ejido* members useful to the community. He envisioned regional variations in the types of families that would be permitted within the ZUE as a result of regional development studies. Agreeing with the Alvarez Vazquez proposal, he also held that in order to keep the land outside the market, residents in the ZUE should only be granted *patrimonio familiar* status.[13]

Covarrubias Gaytán noted that recently introduced agrarian legislation allowing the *ejidos* to engage in some form of commercial exploitation of their natural resources could serve as a precedent to allowing them some freedom to engage in commercial real estate activity not involving transfer of land property rights. Drawing a distinction between the property rights of land and buildings, he suggested that *ejido* enterprises be allowed to engage in leasing and selling buildings for limited time periods, although the land on which they were to be built would always remain *ejido* property. He took as a model the legal figure recognized in Spanish law and in the State of Mexico as *derecho de superficie*, (literally "surface rights") which allows for the separation of building ownership from land ownership.

Skepticism is more than justified on the first part of the Covarrubias proposal in view of the narrow legal possibilities and dismal record of the ZUE form, documented in chapters II, III, and V. However, the latter part of his proposal, granting the *ejidos* the legal right to commercialize buildings on *ejido* land on the basis of *derecho de superficie* legislation, can be seen as a second-best solution to the proposal later presented. It opens up new possibilities for converting rural *ejido* land to urban use. It offers the ejidatarios a legitimate alternative to confiscatory expropriation and an opportunity to participate in the rewards of urbanization. Nevertheless, its applicability as a general conversion formula is still very limited, as it does not respond to the many situations where private property rights of land are required to bring about investment and development.

The other alternative approach that government should take over *ejido* property before urbanization takes place, thus playing the role of main supplier of urbanized land, dominated the conference. It was articulated in various versions by a number of participants such as Antonio Azuela (SMP, 1982: 215-223), Roberto Eibenschutz & coauthors (SMP, 1982: 225), Guillermo Bistrain Reyes (SMP, 1982: 191-194) and again Francisco Covarrubias (SMP, 1982: p. 15-35).[14]

Roberto Eibenschutz, a former undersecretary of SEDUE, and a group of colleagues, looked at expropriation and conversion of *ejido* land within the wider context of government responsibility for creating and managing land reserves for urban expansion as required by the 1976 Constitutional amendment. They envisioned a major role for *ejido* land in supplying the reserve, although land could also be acquired from private sources through purchase, barter and expropriation.

The property rights to the reserve lands, according to this proposal, would not be transferred but would permanently remain in government hands which would only grant usufruct rights to tenants. Such rights would be granted to established corporate organizations such as unions, civil associations, and co-operatives, which would contract with their members as individual tenants. To prevent illegal settlement, these organizations would grant rent-free occupation to the poorest settlers in the initial stage of their settlement process.

Francisco Covarrubias Gaytán, in his alternative proposal, saw local governments as the most fitting recipients of expropriated *ejido* land for the establishment of the reserves. Since municipalities have legal personality with necessary powers to acquire property and to enter into contractual relationships, *ejido* ownership could be transferred to them via presidential expropriation. He suggested, alternatively, that transfer could be accomplished in a more direct manner circumventing the need for presidential intervention as a sequel to the approval process of the urban development plan. *Ejido* property transfer to the municipality in this alternative would take place as a result of the official approval and publication of the municipal urban development plan, prior payment of compensation to the *ejidatarios.*

Covarrubias Gaytán drew inspiration for his proposal from current Spanish law, which requires that ownership of land reserves be vested in the municipalities. Contemporary Spanish law prescribes the mechanisms which govern the management of the municipal land reserves. The land for these reserves is identified according to the urban development plans guiding the expansion of the cities and towns. The land may be obtained by expropriation where necessary. Income derived from the reserves must be applied to their maintenance and expansion.

Guillermo Bistrain Reyes took a more radical statist posture, favoring government appropriation. His view belongs to the statist philosophy of the earlier Echeverria period. He argued in favor of a greatly enhanced role for government in the real estate market to guarantee an adequate supply of urbanized land for urban growth in the different population centers. Unlike other authors of this group, he felt no qualms about government competing for profit in the market, buying and selling land—including expropriated *ejido* land—for every type of commercial purpose. He justified profits by government as a form of income redistribution which would allow government to apply some of its real estate profits to subsidize urban land needs of the lower income population.[15]

Bistrain Reyes proposed that the federal government create a national land system fully empowered legally and financially to acquire, develop, and sell land for urban development. It would be authorized to issue bonds to raise the resources needed to operate. Interest payments on the bonds would be acquired from the system's profits. The system would carry out a variety of functions.

First, it would regulate market land prices through supply of land for housing, commerce, industry and the rest of urban land uses. Second, it would supply the population centers with the necessary land required for their five year reserves, as indicated by their urban development plans. Also, it would supply subsidized urbanized land to the lower income groups. Further, it would subject the use of land to the guidelines of the urban development plans and influence the regional distribution of population and economic growth.

There can be little doubt that proposals such as those just outlined, which advocated government takeover of *ejido* land, were mainly based on the perception of a vast reserve of *ejido* land of great economic and strategic value, readily available in the urban peripheries, that could be obtained at almost no cost by government. Somehow title to this land, which was taken from the *haciendas* for the *ejidos* in the past, was seen as remaining open for whatever public need may arise. The *ejidatarios* were viewed as a minor obstacle—a dependent and docile clientele of the state who could easily be taken care of by the minimal compensation payments established in the agrarian legislation.

The implicit notions that a land reserve necessarily requires government ownership and that government management will assure equitable and efficient allocation of land underlies these proposals. The assumption of government ownership as a sine qua non of the reserve is not only gratuitous, but also ignores a series of far-reaching negative economic and social outcomes. The disruptive effects of government expropriation are well understood in the *ejidos*, as attested to by their reaction to government expropriation. Even Covarrubias Gaytán, who was well aware of the roots of *ejido* opposition, as can be see from his remarks,[16] abstained from drawing the necessary conclusions in this respect.

Private land owner opposition to the implementation of a government-owned reserve could be expected to be even stronger than that of the *ejidatarios*. Even if expropriation were to be carried out at near commercial compensation prices, which is most doubtful, private landowners in the urban peripheries are certain to resent a process which dispossesses them from their property. Threatened owners could be expected to activate defense mechanisms more effectively than the *ejidos* to preempt government expropriation, such as land invasions and illegal subdivisions. The risk of fueling major inflation by acquisition of great quantities of land for the reserve is another most important consequence that would follow from a government-owned reserve. Where land would be acquired long before its use to lower acquisition prices, valuable capital would be tied up unproductively for many years without any return.

The possibility of making all types of land, private and *ejido,* state property was another alternative mentioned in the conference. This course of action, referred to by Alvarez Vázquez in the conference, had been advocated earlier by intellectual radical circles as a panacea to solve land problems in rural and urban areas.[17] It was rejected by most political leaders and legal experts, as Alvarez Vázquez points out, in a wide opinion survey that preceded the IV Reunion of the Republic (1980) presided over by President López Portillo. The negative response from the Mexican political elite, including those representing the worker and peasant sectors, made it clear that such a radical change was not achievable within the institutional framework of the Mexican political system. Furthermore, most prominent Mexican jurists, already alluded to in chapter four, doubted such an action could be accommodated within the 1917 Constitution, and would constitute a revolutionary break with the political system.

The alternative of privatizing *ejido* land in the suburban areas and making the *ejidatarios* equal beneficiaries of urbanization with private owners was offhandedly dismissed. The thought that the *ejidatarios* could participate in the benefits of urbanization brought forth the comment by one conferee: "opening up the agrarian regime to urbanization, would imply the project of converting the *ejidatarios* into urban entrepreneurs, something which does not seem to have any justification."[18] The Constitutional principle that the *ejidos* are owners and not tenants of their land who have rightful claims to participate in the process and benefits of urban development was largely ignored by these conferees in favor of a statist approach.

Some of the papers were critical of the regularization programs, not on grounds of their being antithetical to planned growth, but because of the market effects that might follow when private property titles are granted to the irregular settlers. René Coulomb Bosc, reflecting a current of thinking that is echoed in the literature,[19] criticized title regularization for driving up the cost of living. The increase in value of the land as a result of its newly acquired transferability, the regularization payments, the local taxes assessed on the land, and the higher densities and rents that come with recognized property titles, all contribute to rise in the cost of living. The settlers who can not afford the higher costs may soon be displaced and forced to new settlements in the more distant periphery.[20]

The market effects of regularization seemed to incline some of these participants to join a school of thought which holds that illegality and irregular settlement activity are a lesser evil than regularization.[21] Although they were aware, just the same, of the enormous significance that property titles have for the settlers as the fulfillment of one of their most cherished dream, these participants stated their position that the family properties must be kept outside of the market by legal restrictions to prevent speculation and profit, if regularization is to take place. It can be reasonably assumed that these conferees would have also favored restricting individual property rights and marketability in planned low-income areas established on legally converted *ejido* land.

Coulomb Bosc argued in favor of eliminating individual property in the regularized settlements. He proposed instead collective ownership of the regularized land by housing cooperatives to prevent "speculation and profit." Only organized collective groups should be granted property titles when regularization takes place. He recalled the active struggles of some settler groups and housing cooperatives in this direction and echoed their contention that, "we do not want the individual member to speculate with what was achieved by the efforts and struggle of the whole group."

Others, also wishing to block market effects of regularization suggested instead that the inalienable urban property form known as *patrimonio familiar* be adopted. Joaquín Alvarez Vázquez urged that *patrimonio familiar* status be adopted for regularized family properties on *ejido* land "in order to prevent that the land, once urbanized, should enter the market, and its utilization become dominated by the logic of profit and speculation."[22]

While concern for preventing the negative effects of regularization can be understood, it is difficult to accept their remedies. Fee simple property titles are one of the most cherished goals of the settlers.[23] Limited possession rights that exclude the right to sell and mortgage greatly curtail individual choice and freedom. They go against the wishes of the overwhelming majority of the settlers as proven again and again by the regularization programs.

Proposed restrictions on property rights of the settlers also go against their economic interests. When the properties of a group are isolated from the general market economy, much economic value is lost. While the general market rises, owners of these properties are not legally allowed to benefit from the increases. Restricted in their marketability, these owners transfer or sell in the illegal market at much lower prices. Investment is greatly curtailed; people and properties stagnate while others prosper.

THE UNSOLVED DILEMMA

In the final balance, credit should be given to the Tlaxcala Conference for focusing on the grave consequences of solving the housing needs of the low income population in the periphery of the rapidly expanding Mexican cities through irregular development and regularization. Equally significant was the conference's recognition of the central role played by *ejido* property rights in inducing irregular settlement activity and the urgent need of finding ways of incorporating the *ejido* lands into planned urban growth. However, when one looks at the proposals of the conference, the conclusion must be drawn that no coherent program nor set of changes emerged from the conference to actually facilitate the urban conversion of the *ejidos* within the framework of the planning system and legality, much less on how to use this resource to solve the needs of the low income population as proposed in the international literature.[24]

The conference failed to focus on the roots of *ejido* behavior in the urban peripheries and to grasp the inequities that determine that behavior. Both those who advocated stricter control measures in the *ejidos* in the periphery to stop irregular development and those that favored government massive takeover of the *ejidos* ignored in one way or another the fundamental claims and property rights of the *ejidos* and the consequences of such an approach in a situation of intense urban expansion pressures. The objectives raised by the conference, to find a way to incorporate the *ejido* into planned urban expansion while solving the housing needs of low-income people, have not been met. A policy proposal seeking to meet these objectives will be suggested in the next chapter.

There are many reasons why the *ejidatarios* reject expropriation. These certainly include, but go beyond, their being short-changed by compensation payments assessed on the agricultural value of the land, when its commercial value is incomparably higher. They also include justified fear of the long, protracted,[25] uncertain waiting periods to receive compensation payments, and the agonizing struggle with the bureaucracy. Above all, the *ejidatarios* object to the loss of their means of livelihood, to being side-tracked and excluded from a process in which others reap the benefits. Their reaction to the impending threat of expropriation by CORETT or the reserve described in the San Bernabé and Miguel Hidalgo case studies has been to act on their own and rapidly sell and settle the land illegally, within the constrained limits of the informal market. The words of Milton and Rose Friedman might well describe their plight:[26]

> When the law interferes with people's pursuit of their own values, they will try to find a way around. They will evade the law, they will break the law, or even leave their country. Few of us believe in a moral code that justifies forcing people to give up much of what they produce to finance payments to persons they don't know for purposes they may not approve of. When the law contradicts what most people

> regard as moral and proper, they will break the law—whether the law is enacted in the name of a noble ideal such as equality or in the naked interest of one group at the expense of another.

In spite of the many institutional changes and the efforts to introduce some order in the last twenty years through the planning process, the basic fact is that the *ejidos* have not been incorporated into the process. They remain alienated. Irregular sales of *ejido* lands to irregular settlers have actually increased. It is estimated that in the Cuatitlán Texcoco Valley alone, as many as 700,000 irregular settlers came in last year alone, while in the Chalco valley, intensive irregular settlement activity has already created a major population center.[27] Land use planning and management remains in a powerless condition in Mexico partly as a result of this failure. The land reserve concept has come forward more as a design than a reality. The land use conversion process continues to endanger the environment and is yielding unsatisfactory results at increasingly higher social costs.

The question of *ejido* land conversion in the urban peripheries, so intimately interwoven with the question of the planning system's effectiveness, comes forward as a major challenge that can no longer be avoided. This study and the policy proposal that will follow constitute an acknowledgment of the urgency of responding to that challenge.

---

[1] (DO, February 6, 1976).

[2] (DO, May 26, 1976).

[3] The amendment reads as follow in the original Spanish text: En consecuencia, se dictarán las medidas necesarias para ordenar los asentamientos humanos y establecer adecuadas provisiones, usos, reservas y destinos de tierras, aguas y bosques, a efecto de ejecutar obras públicas y de planear y regular la fundación, conservación, mejoramiento y crecimiento de los centros de población;

[4] (Séptima Epoca, vol. 35, sexta parte. Azuela, 89: p. 61).

[5] (Azuela, 1989: p.50-51).

[6] Ibid.

[7] (D.O. January 7, 1976).

[8] Articulo 115, fraccion 5, Constitución Politica de los E.U.M.

[9] The conference acts were published by SMP. See bibliography (SMP, 1982).

[10] (SMP, 1982: p.165-167).

[11] (Ibid: 37-51).

[12] Patrimonio Familiar is a legal protection granted the poorest families earning below minimum salary, against any claim on their private house or agricultural plot for unpaid debts or obligations. Such property can not be sold, mortgaged nor alienated by any

means.

13 (SMP, 1982: p. 15-35).

14 (SMP, 1982: Azuela, p. 215-223, Eibenschutz & coauthors, p. 225; Guillermo Bistrain Reyes, p. 191-194; Francisco Covarrubias p. 15-35).

15 (SMP, 1982: p. 191-194).

16 "The *ejidatarios* are affected in the expropriation process by the small compensation payments, the protracted administrative procedures for their actual payment, and above all by the loss of the resource that furnished them with a living; it is fundamental that the incorporation of *ejido* land to urban development should contemplate the interests of the *ejidatarios* with social justice" (SMP, 1982: p.16). "Concerning the compensation to the *ejidatarios*, it is not altogether improper to insist on the need to take them into account, not only to determine their role in the process, but also to find a way of optimizing to their benefit the compensation that is owed to them. Generally this last consist in the acquisition of other land for agricultural activity, which may imply a change of location to which not all the *ejidatarios* might be ready. A way is still to be found of integrating them to urban development without hurting their economy" (SMP, 1982: p.34).

17 (SMP, 1982: 37-51).

18 (SMP, Azuela de la Cueva, 1982: p. 215-223). This author argued that the regularization programs were biased in favor of the *ejidos* due to their being more powerful. He maintained that they lost any right to compensation in regularization, on the principle that there can be no rights without corresponding obligations.

19 "There is increasing doubt about the benefits that legalization of land may bring to the urban poor." (Gilbert & Ward, 1985: p. 250-251)

20 (SMP, 1982: 147-154).

21 See the discussion in Gilbert & Ward (1985:p. 250-251, 75).

22 (SMP, 1982; Joaquín Alvarez Vázquez: 37-51).

23 (de la Rosa, 1974; Gilbert & Ward,1985: p. 244; Cornelius, 1975: p. 131, 172-173).

24 (Angel *et al.*, 1983; Varley, 1985:p. 14).

25 The daily press routinely carries reports about ongoing demands for *ejido* expropriations that occurred decades ago.

26 (Friedman, 1980:p. 145).

27 (Excelsior, January 29, 1989).

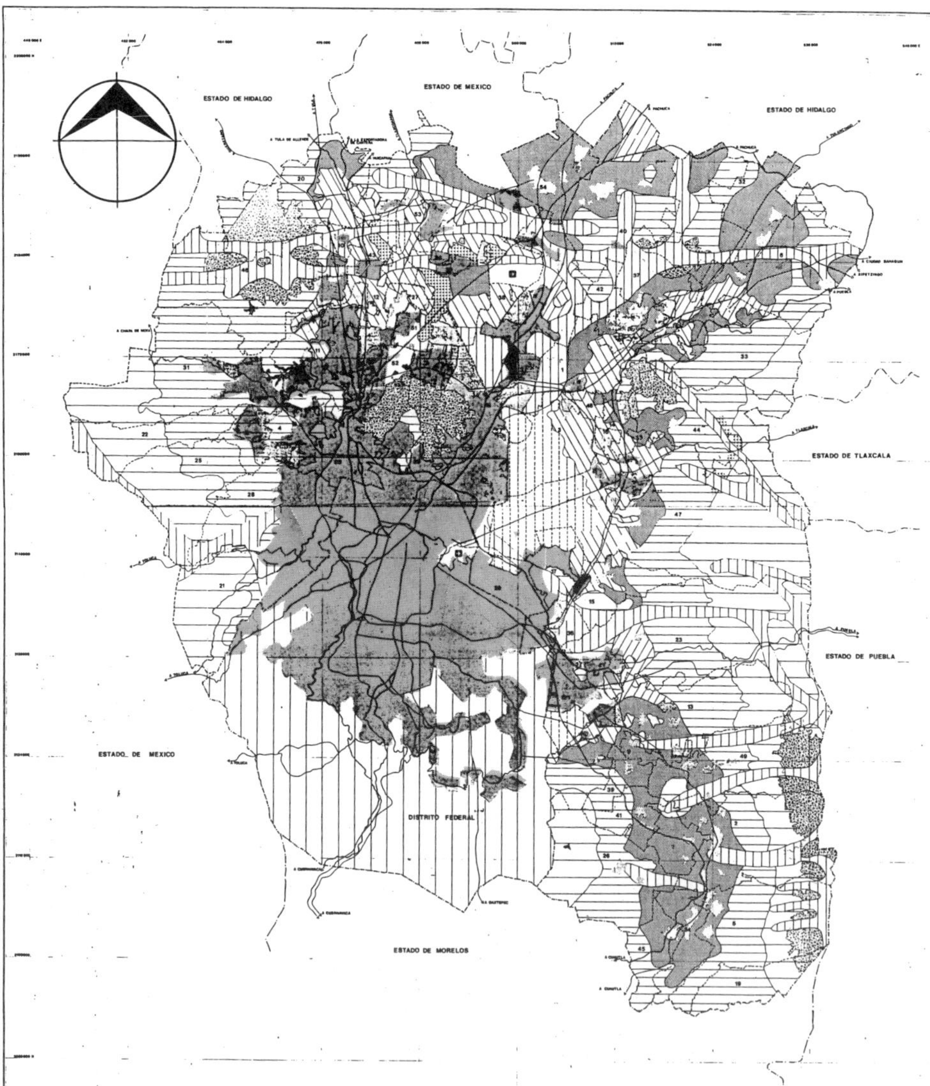

Figure 2. Map of the Metropolitan Region of Mexico City

SYMBOLS

State Borders
Municipal Borders
Main Highways
Railroad Lines
Airport
Urban Area
SM Territoral Reserve
DF Ecological Reserve
Slopes: Between 10-17%
Slopes: Greater than 17%
Flood Risk Area
Seismic Risk Area
Water Body
Irrigated Agricultural Land
Potential Agricultural Use
Potential Grazing Land
Park
Appropriate for Urban Use

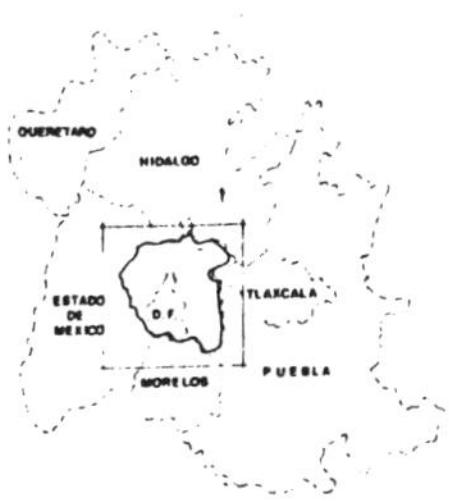

Chapter VII

OUTLINE OF A POLICY PROPOSAL FOR *EJIDO* LAND CONVERSION IN URBAN GROWTH AREAS

AN INTRODUCTION TO THE PROPOSAL

What would be the major ingredients of a better and workable framework that would guide land use conversion and planning policy? To be effective, and workable, the current condition of *ejido* property rights and the related behavior of the *ejidatarios* must be addressed. The program framework must also be structured to enlist support from many or most of the other impacted parties, such as low-income settlers, private landowners, government agency officials, and leaders of the political parties. No new program can be guaranteed support; nevertheless, some chance of acceptability by some groups seems a standard any policy proposal must meet. On the other hand, the mere presence of active opponents to any public policy is not acceptable evidence that the proposed policy would not be highly effective in spite of opposition.

Without suggesting that details of program design can be elaborated upon, recognizing that profound changes by many players and institutions will be required, and that costly set-backs along the way to implementation are certain to occur—a policy proposal will now be presented. The proposal suggests that a set of reliable and feasible actions can be taken to arrive at a more functional arrangement of land uses in the expansion areas, enhancing private and collective well-being. The increase in social and economic value that such actions can bring about would turn the land conversion process into a positive sum-game. Although implementation of this set of actions would involve considerable agency expenses, it should be possible to recapture these costs in the long run through the increased tax incomes that will result from the higher asset values and increased economic rents of the converted lands.

While some of these actions suggested could be implemented independent of land reserves for urban growth mandated in the 1976 constitutional amendment and the Human Settlements Law, the proposal, which intends to facilitate implementation of the reserve, is best understood within the framework of the reserve. A diagram which summarizes the general steps involved in implementing the proposal should be of help in visualizing how it should work within the general framework of the reserve.

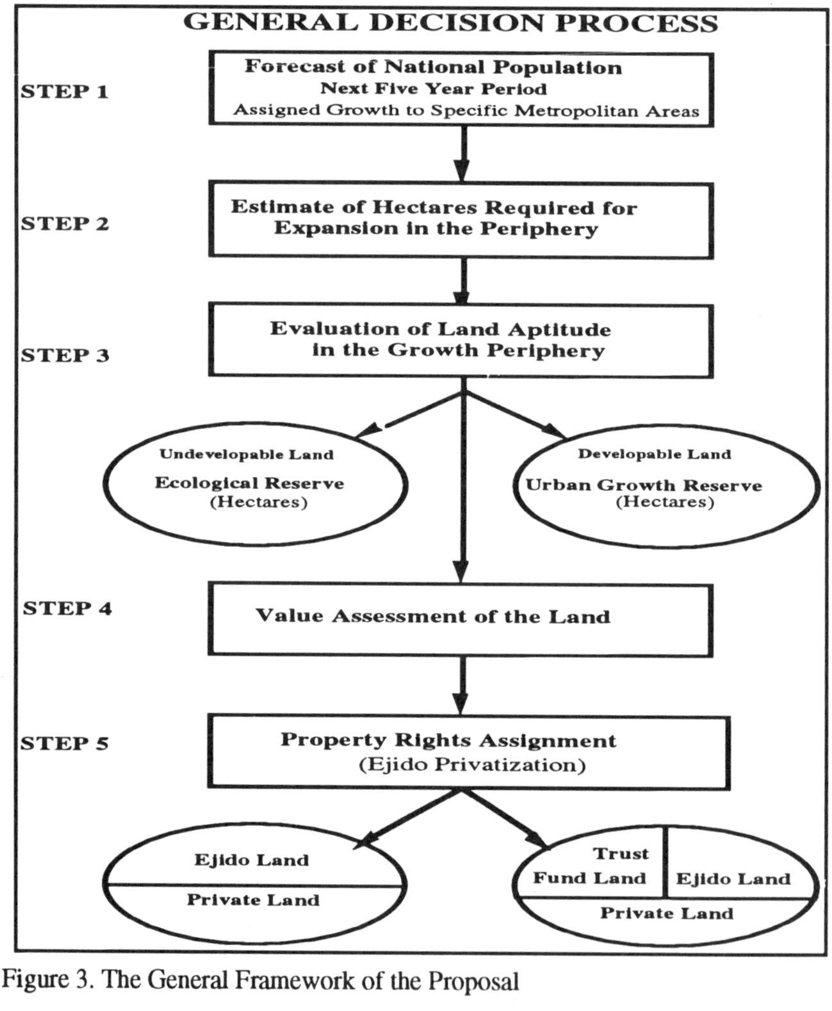

Figure 3. The General Framework of the Proposal

The process graphically depicted in the diagram follows a sequence of five steps. The first step addresses the question of how many people would have to be accommodated in the city and its component sub-areas for the next five-year period; a forecast of the city's population and its component sub-areas should be carried out to provide the answer, using national population models which take into consideration natural and migratory growth as a reference base. The second step would have to determine how much land would be required for the expected population; space standards appropriate to each city would have to be applied to assess the amount of developable land that the reserve should provide. The third step requires that a technical land aptitude study of the potential growth areas contemplated in the master plan should be carried out, to classify the land as developable for the growth reserve, or as undevelopable for the ecological reserve from which urbanization would be diverted. The fourth step concerns land value and requires that value assessment studies should be carried out where necessary for the allocation of property rights between the *ejidos* and the settlement Trust funds. The fifth step would require the actual assignation of the land's property rights between the former holders of the land and settlement Trusteeships.

The proposal that now follows consists of five interrelated sections. Although each is treated separately, their linkages and interaction will become clear in the text.

1) The first section suggests privatization of *ejido* land in the urban peripheries with the *ejidatarios* as the beneficiaries, in step with the requirements of land for urban expansion.

2) The second section suggests that a fraction of the land privatized be allocated for low income settlements.

3) The third section relates the process of *ejido* privatization to the implementation of the land reserve.

4) The fourth section discusses the need to establish a system of judicial review to oversee the implementation of the reserve and discusses the question of compensation claims that may arise.

5) The fifth section examines the way *ejido* privatization would be experienced at the various scales of its implementation, from the lowest scale of the individual *ejidatario* up to the *ejido* mergers.

1. Privatization of *Ejido* Property Rights In Suburban Areas

One of the central ideas of the proposal suggests that privatization of the *ejidos* in the urban peripheries, and not government takeover, should be the route for the conversion of *ejido* land for urban expansion. Privatization of the *ejidos* in the periphery would occur periodically as land is needed by the city for its expansion and protection of its ecological resources. Expropriation in these terms would not require any monetary compensation by government. The return of the majority of the land to the *ejidos* as private property would make any such payments unnecessary and would not entail inflationary pressures on the economy.

To carry out privatization in favor of the *ejidos* it would be required that the beneficiary *ejidos* be given legal powers to act as private corporate entities. The *ejido* corporations would enjoy the same rights and obligations that apply to every private corporate organization under civil corporate law. The *ejido* assembly would be transformed in the privatized *ejido* into a shareholders meeting where major decisions will have to be approved according to clearly stated procedural rules. The individual *ejidatarios* would become equal shareholders in a private corporation not direct owners of individual plots. As equal shareholders, internal resettlement of *ejido* members by the *ejido* corporation within the *ejido* premises could take place as required by the subdivision and development plans, without creating any special property rights problems.

By becoming the private owners of most of their land, the suburban *ejidos* would acquire wealth and the base of credit to develop their resources. The *ejidos* would be in a position to develop business strategies in the formal market that best suit their interests. A variety of options would be available: Lease agreements of their land, sales, and direct development. All legal obstacles to obtain bank financing for any possible development plans on their land would be removed. It would be in their best interest to provide themselves with well prepared professional business plans on how to manage their assets. Public auctions of some of the developable parts of the *ejidos* could be carried out under the supervision of the Banco de Mexico to finance development of the rest of their land.

As a private corporate group, the *ejidatarios* should be free to negotiate partnership arrangements with government bodies or any private enterprise, to develop their land. Such voluntary partnerships are one of the possible ways to make up for the capital and entrepreneurial skills that the *ejidos* are lacking. In such a situation, however, any partner would be required to negotiate fairly based on present and expected commercial values of the land, not its agricultural value. Independent assessment procedures of the land's commercial value, as used by private partnerships, could be adopted to provide the *ejidos* with information on values. Both partners should share the long-term risks and benefits of land development. In no case would the *ejidatarios* relinquish their rights to participate in the decision-making process and the surveillance of their interests.

The *ejidos*, although free to commercialize and develop their land to its fullest economic potential when receiving it back as private property, would nevertheless be restricted by the municipal development plan's limitations, as would any other private owner. As shown in step 3 of figure 3, land at the urban periphery would be analyzed for development suitability before and as a condition for privatization. Significant portions of *ejido* land would be expected to be assigned to the ecological reserve, while others would join the active reserve for urban use. All this land, whether in the ecological or in the active reserve, would become private *ejido* property. Subsequent development and commercialization of the *ejido*, as in the case of any private business corporation, could then take place.

Zoning, the final legal product of the planning process, would have to insure that sufficient land is put aside for productive and commercial activities to bring about an overall equilibrium between jobs and housing. Privatization of *ejido* land in the urban peripheries carried out in these terms would not require a radical change in the agrarian reform law, but could be achieved within the framework of the present law by making the *ejidos*, the present holders of the land, the beneficiaries of the expropriation decrees.

2. Land Allocation to Legal Low Income Settlements

In exchange for privatization and the right to urbanize their land, the *ejidos* would be required to give up a minor fraction of their developable land to municipal trusteeship funds. The Trusteeships would be established in the municipalities where urban expansion takes place, for the purpose of developing legal settlements for the low-income population. Such a transfer of land at that particular juncture is justified on grounds that the higher land values that follow from urban development do not belong to the owners of the land exclusively, but also to society which is a legitimate claimant to part of these windfalls, as Henry George pointed out more than a century ago. Thus a portion of any land, *ejido* or private, should be allocated to low income housing when it is allowed to develop. Although the *ejidos* would be losing a part of their developable rural land, they would be amply compensated by the much higher value of their privatized land. As long as the process opens up their land for net value increases, *ejido* communities would certainly benefit from the change.

The land allocated would initially be transferred at no cost to the municipal Trusteeships as a sequel to the reserve declarations.[1] The Trusteeships would use that land to create well planned, although modestly serviced legal settlements to be incrementally developed for low-income families who are unable to register their demands through regular markets. Managed in the transition process by the state, this land would not become government property, but would eventually become the private property of the settler families. Fee simple title giving the settlers ownership security and full freedom to sell and mortgage, without any undue restrictions, would be granted in exchange for cash payments or sweat equity to develop their community's infrastructure and services.

The proposal seeks an alternative to the irregular settlements, which should help to gradually diminish the need for that dismally inadequate housing solution. Planned, legal settlements, on the fraction of *ejido* land going to the Trusteeships constitute such an alternative. By anticipating the irregular settlers and preparing enough land to develop incremental housing schemes in advance, the government will be in a position to leave behind gradually its surrogate housing policy, based on the promotion and tolerance of irregular settlements and post facto regularization.

Indirect benefits will come to all property owners in the reserve, since many harmful spillovers caused by disadvantaged squatter settlements in the periphery will be greatly reduced. Environmental and efficiency implications may even be greater, because settler initiatives to better their own lives and their communities will be released through property ownership. The contributions settlers must make to gain ultimate and outright transfer of title to the properties assigned to them can help ensure positive settler behaviors. With the promises of their own equity being much higher than in the irregular settlements, their responses toward improvement should come about far more intensely and much sooner.

This initiative is not seen as a panacea to the irregular settlements problem, but rather as an improvement over the present dismal situation. It is perceived as a part of a more comprehensive housing policy for low income people that should be developed over time. The proposal goes beyond the local sites and services programs which are normally very limited in scope, striving to intervene at the more general level of the supply of land where, as rightly pointed out in the literature, "interventions are likely to have far greater repercussions than formal sites and services type schemes."[2]

The proposal draws much inspiration in this respect from the pioneering work of Charles Abrams and the findings of John F.C. Turner[3] many of which were derived from the irregular settlements of Mexico City. Abrams and later Turner established the thesis that solving the problem of land for the low-income people in Third World countries is synonymous with solving their housing needs. If allowed, people will do much of the rest by themselves, even where initially, the land is minimally serviced and the residents have to improve the settlement. Consequently, these authors eschewed building totally finished public housing in favor of supplying the settlers with land, technical assistance, low priced building materials and building licenses.

2.1 The Allocation Formula

To prevent conflicts between the government, owners and settlers the proportion of land to be taken from the developable fraction of the *ejidos* and private properties for the Trusteeships, should be set in the corresponding agrarian and civil laws according to general allocation formulas. Both the amount of land area and the amount of land value would need to be considered.

While it is beyond the scope of this book to join the discussion initiated more than a century ago on the rightful distribution of land development windfalls between owners and society, it is certainly not the intention of this proposal to undermine the economic soundness of land development enterprises. The legislation adopted in the state of New Jersey--now being tested before the American higher courts in the Mount Laurel case--which requires every builder to assign around 20% of his units for low income families, constitutes a salutary precedent providing an acceptable basis for such allocations in a western market society.

Such an allocation when applied to land area as distinguished from building area has the potential of quantitatively minimizing the need for irregular settlements. The actual proportion of land occupied by the irregular settlements in the MAMC in the recent historical past, should provide backing for this contention. Data from the State of Mexico, the main front of irregular settlement activity in the MAMC, show that 28.2% (138.4 sq. km.) of the total amount of land developed from 1940 to 1982 for residential and non-residential purposes in the SM-MAMC (489.2 sq. km.) was used for irregular settlements.[4] A general allocation rate of 20% on developable land can make a decisive contribution in eliminating irregular settlement activity.

Although the principle of allocating a fraction of any land to be developed applies equally to *ejido* as to private land on grounds of fairness, the allocation formula in both cases need not necessarily be identical. While *ejido* property is endowed by the nation at no cost to its beneficiaries, private property is normally acquired by its holders through purchase. This fact places greater social responsibilities on *ejido* property. In any case it is necessary that *ejidos* and private property owners have a voice in determining these formulas and in their application. Hopefully they could participate in the discussions and would have opportunities to challenge outcomes through appeals process.

The objectivity and technical soundness of the valuation procedure is of utmost importance in the allocation process. A clear, standardized procedure which takes into account all relevant physiographic and cultural characteristics of the different parts of a property in assessing their commercial value is required. Equally important in this respect is access to a reliable judicial review process whereby any dissatisfied party can challenge the outcome of the allocation results and demand a public review of the procedure.

2.2 Organization of the Trusteeships

The Ministry of Urban Development, SEDUE and the DF government, whom the Agrarian Reform Law now designates as the recipients of privatized *ejido* land, would carry the responsibility for establishing and administering the Trusteeships. Trust Funds are fully recognized institutions under Mexican law. Many were set up by the Federal government in the seventies to manage privatized *ejido* property. Some of them were established to develop new tourist attraction centers on scenic *ejidos*, and others to regularize illegally converted *ejido* land in urban areas, such as *Fideicomiso* Netzahualcóyotl. The experience already accumulated in organizing and managing resources strengthens the potential effectiveness of the Trusteeships for this task of collaborating with the low income settlers.

Democratic representation of local *ejidatarios*, settlers and private owners should be guaranteed on the governing boards of the municipal Trust Funds. The Trust Funds would be accountable to the settlers and the Ministry of Urban Development and to the SPP for all their transactions. Periodic audits of their operations and assets would have to be carried out and made public to maintain standards of public ethics. Audits would help assure that all the land entrusted to the Funds every quinquennium would actually be distributed to settlers, and that any land held for commercial/industrial purposes to benefit the settlements is properly managed.

A variety of skills would be needed by the staff of the Trusteeships to carry out their responsibilities: legal, managerial, and technical. Many of the necessary technical skills could be provided by the Ministry of Human Settlements, the DF government, and the public housing organizations through long-term official agreements which would bind them to provide technical services at the request of the Trust Funds. An internal government budgeting system would compensate these organizations for the services that they would be rendering to the Trust Funds. Executive responsibility, however, should lie with the Trust Fund's own limited, highly skilled permanent technical staff, which should excel in community organization and self-help promotion.

The Trusteeships would possess the full legal capacity to contract with groups, individuals and enterprises, issue private property titles, and actively assist settlers in developing land. They would draw up specific legal sale contracts with each settler family by which the terms of their payments in sweat equity or in cash would be clearly spelled out. Down payments when receiving the plot would generally have to be eliminated. Property titles would be released to the settlers after the agreed payments of labor or money would be completed. In the meantime, if a settler family wished to move away from the settlement, resale or barter of the property would have to be carried out through the Trusteeship Funds, in order to discourage "professional" settlers from discrediting the distribution process.

Distribution of the lots in the legal settlements would require clear allocation principles, such as a maximum limit of one lot per family and actual occupation of the lot by every aspiring settler family, excluding occupation by proxy. The selection of applicants would be based on official clearly established eligibility criteria. The purpose of the criteria would be to assure that the actual goals of the program are achieved, not to impose unjustified limitations. Family status and size, maximum allowable family income and a clean police record can be seen as the basic criteria. Some additional eligibility criteria might be considered if necessary, such as exclusion of single households, and a minimum period of local residency in the city prior to the application.

The selection process should be open to public scrutiny. After passing a standardized procedure to ascertain their deserving status, aspiring settlers would enter an official computerized waiting list as individuals or as part of a group. Appeal should be available for applicants rejected as deserving settlers. SEDUE, the ministry responsible for the process, would issue every applicant an untransferable officially numbered certificate recognizing them as deserving settlers, stating the date of issue and geographic preferences. Strict adherence to seniority should be a basic principle of the distribution process. A lottery could be used to choose among applicants with similar seniority, where enough land is momentarily not available for all.

2.3 Role of Government and Settler Autonomy and Initiative

Endowing the settlements with infrastructure and services, and adequate design, is, as a matter of principle, a fundamental responsibility of government. Responsible government cannot evade this basic public duty. The government's random and belated introduction of services in the irregular settlements should be substituted by a more enlightened policy which assumes responsibility for the low-income settlements from their very beginning. Making available serviced land with property title for low-income people to build their own housing is far more effective than building finished housing as far as the number of beneficiaries is concerned, considering the great cost difference per unit between the two alternatives.

Provision of infrastructure and services for the settlements is envisioned as an incremental process. The settlements would start from a very modest infrastructure and service base, which should gradually be upgraded over time as the resources of government, credits, the Trust Funds and the settlers would allow. Timetables for infrastructure building and service introduction, as well as blueprints of their final form, would guide the Trust Funds and the settlers in their improvement activity.

Once the land is legally transferred with clear private title to the Trusteeship Funds, the public housing organizations would be called in to assist the settlers in the general design of the settlements and their individual housing. An open-ended, long term relationship would be established between the local government, the Trust Funds, the housing organizations, and the settlers. Trust Fund personnel specialized in community development would organize the settlers to carry out the community development programs, within the terms established in their sales contracts.

As crucial as the role of government can be in building the settlements' infrastructure and providing services, the improvement process in the legal settlements could, paradoxically, be much accelerated by freeing them from their absolute dependence on government service monopolies. Autonomy and self-help should not only be encouraged and assisted in building housing, but also in building infrastructure and services. In no way implying that the legal settlements should forgo government assistance, granting them legal autonomy for services provision and infrastructure building through the Trusteeships, and outside of them, as a complementary alternative to government provision, represents an added opportunity for improvement in the settlements.

Instead of being limited in their ability to access every service, such as street paving or public transportation through the government, the legal settlements would gain much from being empowered to carry out their own development initiatives where possible. Community organizations established or recognized by the Trust funds should be allowed to help themselves in running their own taxi service, installing their own water or power network, and paving their streets, as advocated by the international development agencies. They should be free to contract with private utility companies which could provide services. Public-private partnerships are, of course, another alternative that should also be available and encouraged.

2.4 Sources of Funding for Infrastructure Development

Although there is no intention to enter into a detailed discussion of the sources of funding to develop the legal settlements, it is clear that considerable amounts of capital and credit will be required over time to provide infrastructure and services for the legal settlements. A variety of such sources can be envisioned.

Considering the settlers initial minimal payments capacity, government investment and soft credits will continue to be the main source where resources for developing the settlements will come from. Loans, national and international for developing infrastructure and housing have been a major funding source in recent decades in Third World countries for improving the irregular settlements. By acting as a catalyst for such credits, government can make a major contribution to developing the settlements.

However, much as the legal settlements will require external assistance the ultimate long-term goal is development of the settlements' own economic capacity. The settlers payments in cash or sweat equity labor as already been mentioned as an important economic resource in this respect. Private utility companies can be seen as another important source of credit. The settlements should be empowered to contract directly with such companies the credits for their infrastructure and services. The precedent of low income communities which pay back their running water nets in short periods, from two to five years, should support the feasibility of this most important private credit source.

The land managed by the Trust Funds can become another important internal source. The settlements will require commercial services and jobs for their population, which can be developed on a fraction of the land held by the Trust Funds. By leasing or selling the land on which these services would be established, or by direct entrepreneurship or profit sharing in them, substantial income can be obtained for developing the settlements infrastructure. Taking as a reference the net percentage of land that is actually used for industrial/commercial activities in the SM-MAMC as a whole (Table 11, Ch. III), 15% of the land held by the Trust Funds could be assigned for this purpose without undermining but actually supplementing the basic function of the settlements, which is that of housing low income people. Effectively managing this resource, the Trusteeship Funds could obtain a cash flow for building basic infrastructure, such as water, sewers and roads, while at the same time providing the community with commercial services and employment opportunities.

Property taxes as a long term source of income must be carefully nurtured from the outset of the settlements. Such a policy would have to be clearly related to the level of services provided. focused in the early stage of the settlement on educating the settlers to the principle of paying taxes and its direct relationship to service provision, not on producing important resources. The vicious circle that Mexican cities have been unable to break, which keeps many millions of urban residents completely outside of the tax base, mainly as result of their irregular tenure situation, has stood in the way of providing them with the necessary infrastructure and services.[5] The clear property rights situation, which would be established in the legal settlements from their very beginning with fee simple property titles for each settler family, would finally allow to break the iron grip of acute chronic fiscal insufficiency.

A long-term tax policy for the legal settlements would have to be designed, considering the minimal capacity of the settlers to pay taxes in the initial stage of their settlement. Only gradually would these taxes approach more realistic assessment values as the settlement is consolidated. The tax/services policy would stress growing individual responsibility for the collective good and advancement to mature citizenship, and would require systematic public communication on the subject. The policy should help prevent that extraction of payments by corrupt local leaders, which occurs in the irregular settlements, should take hold in the Trusteeship settlements.

The administrative framework for property taxation here described would require that the settlers be provided in their tax bills with information on the subsidies that they are obtaining. The differential between the assessed real tax value and the actual tax would have to be made explicit to each taxpayer in his tax bill.[6] The gap between these two values would diminish with time, as the settlers' capacity to pay rises.

2.5 Design of the Settlements

The design of the settlements should offer a variety of options beyond the detached one-family lot. The three-and-four family lot, which allows for a more intensive use of land, should also be made available. The different subdivision options considered will generally have to be compatible with self-help construction, which is seen as the dominant form of building that will prevail in the settlements. Self-help is not to be identified exclusively with traditional workmanship, it can also be made compatible with the use of industrially-produced parts and even prefabricated building systems by individual or groups of households. The industrially produced kitchen/bathroom units have an important role to play as the initial nucleus around which the settler house can develop incrementally over time. Mexico, as many other countries over the world, has shown interest in combining self-help construction with industrial technology. The accumulated experience should be of great value in developing the settlements.

2.6 The *Ejidos* as Legal Settlement Promoters

In addition to the establishment of legal settlements by the trust funds, the proposal also envisions a market-oriented strategy to encourage legal low-income developments on the privatized part of some *ejidos*. Although much of the privatized land of the *ejidos* would be in demand for higher income residential purposes and industry if the proposal is implemented, they would often, as in the past, still find it profitable to supply land for the lower middle class in suburban areas that do not attract upper class residents. Much of the northern and eastern periphery of the MAMC to which upper class developments have not been attracted has the potential to play such a positive future role.

Great profits were made by companies that promoted irregular settlements in the eastern and northern periphery of the MAMC. Private capital investments from national and even foreign interests were involved in their commercialization on a grand scale, attracted by the high rates of return of these ventures.[7] The profit potential that Netzahualcóyotl and other irregular company-sponsored developments have proven, support the concept that the *ejidatarios* could participate in supplying a legal market alternative for many irregular settlement residents. If the necessary steps are taken to encourage and guide this trend within the framework of legality and the planning system, the privatized *ejidos* could play a positive role in creating a market alternative for the lower middle class, who, for lack of a better option, have been forced into the irregular settlements.

The literature as well as the official studies on income distribution has made clear that the residents of the irregular settlements—which now comprise over 60% of the total population of the MAMC, a proportion without parallel even in the Third World—are not a homogeneous group of people with a uniform level of income and housing potential. The 1980 income data of typical irregular settlement municipalities here included, such as Netzahualcóyotl, Chimalhuacán, and La Paz-Los Reyes in the State of Mexico, show that individual income differences existing within their populations are not radically from that of the MAMC as a whole (table 17).

Table 17. Individual Income Distribution by Minimum Wage Groups in the SM-MAMC (1980)

Income Group	Percentage of Population in Each Income Group			
% Minimum Wage	SM-MAMC	Netza	Chimalhuacán	Los Reyes
0 to 100 %	37.8 %	39.1 %	34.0 %	41.8 %
100 to 200 %	40.5 %	47.6 %	45.5 %	47.3 %
200 to 300 %	8.8 %	7.4 %	10.2 %	6.0 %
300 to 400 %	4.4 %	3.0 %	4.6 %	2.5 %
400 to 600 %	3.8 %	1.4 %	3.3 %	1.3 %
> 600 %	4.7 %	1.4 %	2.3 %	1.0 %
Total	100.0 %	100.0%	100.0 %	100.0 %

People/Household 5.0
Earners/Household 2.6
Official Daily Minimum Wage: $MP 163.0
Conversion rate: $US 1D/$MP 22.8

Source: B. Dubost, Suelo Urbano- Estado de Mexico, 1981, Table 7. q. Iracheta

Various income strata can be distinguished in very broad terms in the irregular settlements. The existence of these strata becomes evident to any external observer visiting the more consolidated irregular settlements, where he/she will often find some relatively well-built houses, private cars, and other material signs of a higher living standard against a dominant gray background of shacks and a minimal living standard. The economic stratification of the residents is the result of differences in individual income and in the number of wage earners per family (table 17).

The lowest strata earning below the minimum wage stand below the "absolute" level of poverty; they struggle for mere survival and have more urgent concerns than better housing. This group cannot aspire today to even the most modest form of property in the farthest periphery. They live in crowded one-room shacks, very often paying a small rent to the untitled owner of the lot. As already pointed out by some critics, they may even be harmed by the regularization and improvement programs which raise the cost of living in the irregular settlements.[8] The next higher group earning between one and twice the minimum wage also lacks sufficient income capacity to solve its housing demands through the market. Nevertheless, the two lowest groups are the ones which have benefited the least from the government housing programs, although they are the ones most in need of assistance. They have been left out from the housing programs of the public agencies, while the higher income groups which have a significant market potential by themselves have become the subjects of these programs.

While there are too many variables to draw an absolute line in the income statistics to define the lower middle class, it would not be wrong to suggest that the strata with incomes above two times the minimum wage could benefit most from such market-oriented legal *ejido* settlements. This group includes important numbers of blue and white-collar employees with a steady income, and a majority of self-employed struggling individuals with an unsteady income, who occasionally receive some additional cash, which they might wish to save or apply to their housing needs.

The families in these strata seeking suburban family lots instead of finished houses as discussed in chapter II, have been forced into the irregular settlements for lack of a more adequate housing alternative. They are in the irregular settlements more because of faulty institutional and social arrangements than to acute buying power limitations.[9] The only market alternative available for most of them are the irregular settlements such as Ciudad Netzahualcóyotl. There, uncertain property rights, use of land totally unfit for urban use, and collusion of private and public interests combined to create for a population of over a million people, "the most mechanized, economically and socially exploitative example of urbanization and urban design that one could find in Latin America." [10]

Various means are suggested to encourage the participation of the *ejidos* in the commercial development of their land for legal settlements for the lower middle class. Incentives can be provided to encourage and guide *ejido* participation in this market, such as additional compensatory development rights where the land is used totally or partially for lower middle class settlements; special land development regulations allowing an incremental process of infrastructure building and service installation; official assistance in design and in procuring official subdivision approval; and loans and subsidies for infrastructure and the introduction of services. These incentives should be balanced by an effective consumer-protection apparatus, an area in which Mexico has made great strides in recent years, and by conditional use of the land for legal, well-designed, lower middle class settlements.

3. The Proposal and the Urban Land Reserve

The proposal's suggestions have an important bearing on implementation of the urban land reserve, on which the discussion will now focus. A diagram (fig. 4) which synthesizes graphically the process leading to the establishment of the reserve should prove useful to the reader as a guide for envisioning the reserve concept and the policy proposal's suggestions on the reserve which are developed in the ensuing text.

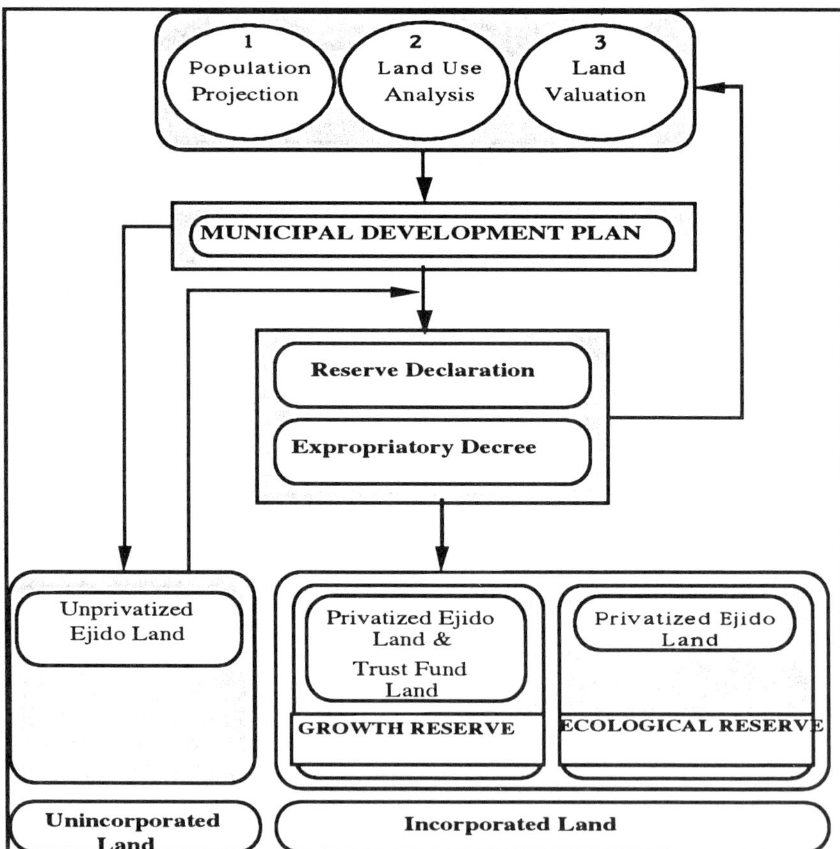

Figure 4. Land Distribution System of the Reserve

The proposal's suggestions have an important bearing on implementation of the urban land reserve, on which the discussion will now focus. The reserve envisioned in the Human Settlements Law as a planned alternative to chaotic irregular expansion, would provide a city with sufficient land for expansion and guide the location and time of the conversion of land. According to the Human Settlements Law, the reserve should be updated every five years by means of periodic reserve declarations. The declarations should define the expansion areas of a city and exclude urban development on ecologically sensitive lands that should not be developed. Zoning plans and detailed land use planning schemes subsequently complement the declarations to define specific land uses, such as residential, commercial, industrial, or others.

The Human Settlements Law indicates that the reserve declarations should be based on the municipal development plans *(Planes Municipales de Desarrollo Urbano)*, which have the more general function of guiding the total process of urban development. The municipal development plans are designed to lay down the long term objectives of the urban expansion process and indicate the physical form of the city. Once the development plan—which can be periodically revised—is approved by the state legislature, it becomes the general basis on which that body can issue every five years, short-term reserve declarations that pinpoint on official maps the specific areas where urban growth should be allowed or excluded in the coming five-year period. Obviously, the long-term plan and the reserve declaration have major judicial and economic significance and implications.

The different space and time frames of these two planning tools should be recognized. The plans have the wider objective of guiding urban change in the total city, including the existing urbanized areas, while the reserve declarations are only intended to guide urban development on the expanding periphery. The time frame of the municipal development plans is also much wider. The municipal development plans establish long term objectives while the reserve declarations have a limited time horizon of five years.

The reserve declarations are not conceived as static replications of the municipal development plan but are intended to periodically rectify the plan's assumptions and results. They also determine the sequence in which the totality of the developable area foreseen in the municipal development plan is to be incorporated into the city. Until a specific five year declaration incorporates peripheral land as developable to the growth reserve, land remains undevelopable and part of the ecological reserve.

Forecasts of expected total population growth, migration included, are used to determine future space needs in each city. National demographic growth models are the proper instrument for this task. The population forecasts can then be translated into space needs by means of human land demand and density standards for each particular city. Metropolitan area space estimates must then be broken down into metropolitan quadrants or sectors.

To determine the areas to be developed or permanently left in their natural state, the municipal development plan and the reserve declarations require a land aptitude analysis. That analysis seeks to identify the environmentally sensitive lands, such as wetlands, aquifer recharge areas, and forests, or archaeological and historical sites, which are to be protected and kept outside of urban development. Sites that are at high risk from natural disasters such as flood plains, desiccated lakes, dry river beds, unstable mountain slopes, soft soils, geologic faults, collapsible caves, mines, and garbage dumps that threaten human life and property would be identified and excluded from development. The same is true for areas exposed to technological risk originating from the installation of dangerous pipelines, deposits, and industrial plants.

Once identified, all sensitive land sites not to be developed would be defined and assigned to the ecological reserve.[11] Some high risk areas could become eligible for development—but only when a special justified need arises—on the condition that special stringent building standards are applied, or protective works are carried out to reduce their risks. Either of these last two alternatives, however, imply high public or private costs which should normally be avoided.

The reserve declarations are not simply intended to reflect the classification of the peripheral lands according to their ecological and risk characteristics in a mechanical manner. They are also designed to induce whatever physical form or model is envisioned for a city in its municipal development plan.[12] The physical planning model is normally the result of combining environmental concerns, such as nearness of the built-up areas to nature, with such other objectives as achieving a rational relationship between employment and housing areas, or optimization of the public transport system and infrastructure costs, as illustrated by such a well-known contemporary development plan as that of Washington DC. Among various other limitations, maximum density restrictions are one of the main instruments by which these objectives are achieved.

While identification of the conservation and vulnerable areas to be excluded from development is an objective process in which technical-scientific methodology prevails over subjective judgment, the decision on the urban form desired as previously described and the sequence in which developable land should be included in each new reserve declaration to achieve this form, implies a more subjective and less deterministic decision process, which depends greatly on the criteria chosen. In this final stage the procedure loses much of its objectivity and can become subject to a power struggle among competing interests.

A democratic planning process in which the community can participate in making the decision on the specific physical expansion model to be adopted by the municipal development plan, as well as on the sequence in which developable land is to be incorporated into the reserve declarations, lends legitimacy to the results. The final approval by a democratically chosen body, such as the legislature, which is capable of publicly scrutinizing the total procedure, and of accommodating differing criteria, conflicting interests, and dissent on these crucial issues, before reaching a binding decision stands on far more solid ground than expedient approval by an autocratic bureaucracy.

Valuation of the different areas to be included within the reserve as developable or ecological land is another important component which feeds the analysis and decision process of the municipal development plans and the reserve declarations. Estimates of the value of the different lands in the periphery are necessary in carrying out the total land conversion process, including expropriation of the *ejidos* and partition of the their land between the *ejidatarios* and the settlement Trust funds. The valuation procedure would apply physiographic and cultural criteria among others, such as location, accessibility, slopes, drainage, and present level of infrastructure and services, to determine the value of the different parts.

Having introduced the reserve concept, discussion will now center on the way the policy proposal meshes with the implementation of the reserve. The proposal turns away from the general approach, rarely questioned, which assumes that the reserve requires government ownership of the land. This proposal rejects that approach and adopts the position that government takeover and ownership of the growth reserve lands is unnecessary, except of course, where the government expropriates for specific public interest purposes.

The proposal envisions the reserve as a government regulatory mechanism, not a government monopoly. In consequence there is no need for the government to become owner of the *ejidos* or private land holdings to implement the reserve; to the contrary, the *ejidos* would be privatized at the time that the reserve declarations are issued. The long-term development plan and the reserve declarations following would bring about a process in which the land of the *ejidos* in the urban periphery is privatized according to need, and placed according to its characteristics within the growth reserve or the ecological reserve.

The proposal envisions a process of *ejido* land privatization closely co-ordinated with the issuance of the reserve declarations. Before a new declaration is to be issued, the agrarian authorities would be required to synchronize the privatization procedure of the *ejido* areas involved with the proclamation of the reserve declaration by the planning bodies. Presidential expropriation of *ejido* land could remain if considered necessary the end of the process by which *ejido* conversion would be carried out, but it would be in favor of the *ejidos*, not government.

The privatized *ejidos* would join the other private owners within the urban growth reserve in the venture of urban subdivision of their developable land, while their undevelopable land entering the ecological reserve would be left with them as private property for continued agricultural production or other compatible activities. Basic property rights of land would not be altered within the ecological reserve, and private owners and privatized *ejidos* would be free to sell, buy, and lease land, as long as they complied with its restrictions. Land placed within the ecological reserve should normally not be prevented from being used by its owners for a variety of income-producing activities such as agriculture, timber harvesting, the raising of livestock, extraction of minerals, and open air recreation. Golf courses and very low density tourist installations would often be considered acceptable uses.

To protect the ecological reserve against illegal urbanization, a government surveillance and control system would have to be established. Legislation would grant the system ample powers to apply all the incentives and sanctions necessary, including the capacity to resettle squatters and to prosecute violators. The *ejidos* would become most important participants in the surveillance system together with the federal (SRA, SEDUE) and state governments (DDF, SM). Special surveillance committees would have to be set up in each *ejido*. The greatly increased rights of the *ejidos* as private owners of their land would be linked to greater responsibilities as custodians of the ecological reserve.

The aims and practices of the regularization programs would also have to be reexamined. A permissive regularization policy granting recognition to illegal occupation of ecological reserve land would run counter to the protection of these lands. A clear public awareness that neither urban property titles nor

public services will be obtainable in such areas is necessary. Regularization would have to be confined to areas which were irregularly settled before the new policy was implemented, while resettlement would be the policy to be followed in new squatter areas on ecological reserve lands. Political will-power and strict government adherence to its policies would make sure that new Netzahualcóyotls on ecological reserve lands will not rise again.

For the reserve concept to actually work and guide the expansion process within the assigned growth areas, government would have to free enough *ejido* and private land through the reserve declarations to at least cover the growth needs of the different population sectors for the next five-year period. Failure in this respect can turn the reserve into an oppressive rationing system where only those with economic or political power can gain access. Urban expansion in such a scenario would not be halted, but would seek escape valves, as disruptive as in the past, in irregular development outside of the assigned areas. The planning system will therefore have to closely monitor demographic and economic growth trends to detect deviations in population and space needs projections. It is suggested that a revision of the declaration should be required in the middle of each five-year period to remedy unforeseen shortages.

The legal settlements program would have to be targeted to low-income people. If land in the legal settlements is diverted from the intended low income population or used to alleviate high densities of existing areas, demand for illegally converted *ejido* land could be expected to surface again. Elimination of that demand would consequently depend not only the availability of enough privatized land for current growth needs, but also on targeting the low income settlers wherever they are, as previously indicated in the proposal.

4. Reserve Judicial Review and Compensation Claims

Some judicial problems may arise with implementation of the reserve. These problems are not derived from the policy proposal, but are rather inherent to land use controls in general, and are being experienced in every western country where these controls are applied. Land use control devices, such as the reserve or zoning, may affect some particular interests; some owners may suffer losses when these controls are implemented, bringing about claims by the aggrieved parties. These claims have not undermined land use controls, and the courts have upheld their legitimacy on the principle that the public interest takes precedence over particular interests. Many unsolved puzzles, however, remain concerning the right to compensation in the western countries where these land use controls have been applied. Conflicting views on the issue of compensation have often been expressed by the courts within each of these countries.

Implementation of the reserve in Mexican cities can be expected to bring about judicial conflicts and claims by those who might feel that their interests are hurt. A judicial review system to manage such conflicts is therefore necessary, whereby all types of claims, including compensation claims, could be heard and taken care of, as with zoning in other countries. Claimants should be given a legitimate opportunity to appeal decisions and actions that they consider unfair or unwarranted. The system should oversee the total process of the reserve and have the necessary legal powers to redress inequities.

Considering the degree of specialization that would be necessary to handle claims derived from implementing the reserve and the number of cases to be handled, it might be advisable to set up special courts for this purpose. Case law standards will gradually emerge from them which will diminish uncertainty costs. Another most important complementary alternative to be considered is the establishment of panels to settle disputes which could negotiate voluntary agreements outside of the courts. Participation of local citizens in peer juries should create greater trust and provide first-hand knowledge on the parties and issues involved.

Disagreement with the procedures used in classifying the land as developable or undevelopable can be expected to become an important source of claims by *ejidos* and private owners. The judicial review system should have the powers and capabilities to determine whether the procedures that would be used lacked the necessary technical basis or were biased in favor of special interests. That would provide checks and balances against domination of the process by special interests. Although such a judicial review system might be perceived as slowing implementation of the reserve and weakening its expediency, it will actually strengthen trust in the system and insure its long range stability.

Assessment of land value can also become a potential source of conflicts and claims. Land value assessments play a most important role in the partition of land between the *ejidos* and the housing trust funds. Judicial review of the assessment procedures can make an important contribution in guaranteeing their proper implementation.

Aside from claims against flawed land classifications and value assessments, one could expect claims of compensation due to loss of opportunities. Implementation of the reserve, as previously explained, can be expected to activate compensation claims by the *ejidos* and private owners whose rural land is assigned permanently to the ecological reserve. While at present no compensation is paid to owners whose land is included in the ecological reserve, as is the case with the DF southern ecological reserve, the courts will have to decide on the constitutional basis of compensation claims and face the issue of whether inclusion of land within the ecological reserve constitutes a taking.

The first constitutional question to be resolved in this case is whether urban development rights are part of the original bundle of property rights that the Mexican state has granted to private or corporate owners according to article 127 of the Mexican Constitution of 1917, or whether these were retained by the state. If development rights remain the sole prerogative of the Mexican state and are not part of the original bundle of rights that were ceded to private owners, such claims for compensation, when land is excluded from development by the reserve declaration, would lack a constitutional basis. However, if development rights are part of the bundle of property rights originally ceded by the nation to private owners, then implementation of the reserve may lead to a recognition that such claims are indeed justified.

The next question then to consider would be whether the inclusion of land within the ecological reserve constitutes a taking. Arguments on both sides of this issue can be visualized. The argument that a taking occurs would point to the fact that owners of land included in the ecological reserve were being denied the opportunity to use the land for urban development purposes and to benefit from the increases in land value linked to development. On the contrary side, it could be argued that basic property rights would remain and private owners would be free to buy, sell and lease land within the reserve, benefiting from increases in its agricultural commercial value. As owners, they could use their land in a variety of income-producing activities, such as agriculture, timber harvesting, livestock raising, extraction of minerals and recreation, including golf courses and low-density tourism installations. Last but not least, they would be exempted from the tax increases that owners of developable land would be required to pay for developing urban infrastructure in the region.

Assuming that the right to compensation is recognized for owners of land included within the ecological reserve, the next question to consider would be on how to compensate the aggrieved parties. Different alternatives can be visualized. Considering that the reserve is established for the general public good, and that public benefits are derived for the city as a whole, compensation by the state may be one possibility. Such compensation may take a variety of monetary or non-monetary forms, such as property tax advantages for owners within the ecological reserve. Another possibility could be profit-sharing by owners included in the urban reserve with those assigned to the ecological reserve.

Profit sharing by owners in a municipality or in a more limited geographic area is a formula which has been applied outside of compulsory legal compensation situations. It has been used voluntarily in some regional development schemes by groups of neighboring owners convinced of the need of protecting the quality of their environment, such as in the case of the Brandywine valley in Maryland.[13] The owners have achieved this, by merging their properties into a share holding corporation.

The concept of profit sharing could be applied as well to the component areas of the reserve, to secure an equitable form of development and overcome possible opposition from those *ejidos* who may see themselves as losers when much of their land is assigned to the ecological reserve. Legislation could incorporate this possibility into the Human Settlements Law, regardless of the judicial outcome on compulsory compensation. *Ejidos* included in a reserve declaration within a municipality or a smaller geographic unit could merge into a consolidated corporation as part of the conversion procedure. The shares that each participating *ejido* would obtain in any such merger would depend on the assessed commercial value of its land without development restrictions (fig. 3). This procedure would guarantee that every *ejido* receives a fair share in the total municipal equity, eliminating the need for further claims to compensation. Each *ejido* would have its total equity proportionately distributed in privatized marketable land and in non-marketable *ejido* land. Even if a particular *ejido* lay entirely within the ecological reserve, it would still have its proportional share in the merger, in proportion to the commercially assessed value of its land.

Once the conversion procedure has been carried out, the consolidated *ejido* corporation would then be free to divide into smaller corporate groups and apportion land among them according to their shares. Internal feuds among *ejidatarios* make it necessary that an option of partition should always be available once the merger has been carried out. It would also allow in these situations for money compensation to be exchanged for land, where an equitable partition cannot be carried out, or where the *ejidatarios* are willing to trade privatized property rights. All these situations can be handled under the same business and legal principles that apply to the merger and breakup of share holding corporations and partnerships.

5. The Different Scales of *Ejido* Privatization

The proposal will now examine the way *ejido* privatization would be experienced at the various scales of its implementation, beginning at the lowest scale of the individual *ejidatario* up to the *ejido* mergers.

For the individual *ejidatario*, privatization of a suburban *ejido* would represent a change from limited marketable rights over his agricultural parcel within his *ejido*, to fully marketable shares in a land holding corporation in which he is an equal partner with all the other *ejido* members. He will have to learn the meaning of being a shareholder, and the rights, responsibilities and risks involved. Such a process may be fraught with difficulties, particularly where crucial decisions which he may not fully comprehend or approve of are adopted by the *ejido* executive leadership. The individual *ejidatario* will experience the difficulties and problems of the conversion process of the *ejido* from an agricultural community to a corporation or cooperative commercializing suburban land.

Situations may arise where the interests of individual *ejidatarios* may come in conflict with those of the *ejido* enterprise. For example, a situation may arise in which it would be necessary to relocate some *ejidatarios* from their former sites. *Ejidatarios* living on the Trust Fund fraction might be required to leave their sites when the land is partitioned between the Trust Fund and the privatized *ejido*, unless special arrangements are made for them to remain there. Similar situations would occur whenever the *ejido* assembly would decide to commercialize or develop sections of its own privatized land on which individual *ejidatarios* might be living. It would also arise on *ejido* land assigned to the ecological reserve where the *ejido* assembly would decide to sell that land or to change its use within the limitations established by the reserve declaration. The *ejido* assembly would face the problem of resettling the displaced *ejidatarios* in a manner that would be acceptable to them, or of finding other ways of compensating them.

Ascending in scale from the individual *ejidatario* to the *ejido* community as a whole, privatization would constitute a change from a static, rural commune living on non-marketable agricultural land to a suburban land holding corporation possessing marketable urban property rising in value. The transition would present a series of challenges to the *ejido* community for which it may not always be prepared, but with which it will have to learn to cope. To operate within the urban reserve is one of them.

The reserve declaration would classify the land of the *ejido* as developable or undevelopable. In the particular case of a larger *ejido*, only a part of which is incorporated in the reserve declaration, such an *ejido* might find itself controlling unmarketable collective land in addition to privatized ecological reserve land and privatized developable land. Land placed in the developable category would rise greatly in value although the rise would be partially offset by the allocation of a fraction of that land to the Trust Funds. Land placed in the ecological undevelopable category would rise less in value, but the *ejido* would retain the totality of that land, which it would be free to sell.

The partition of the developable part of the *ejido* between the *ejido* and the Trust Fund would involve negotiations. The *ejido* would have the right to choose the parts that it would like to retain within its part. The *ejido* might find that it does not agree with the official assessment value, and would have a right to appeal to the judicial review system or accept arbitration. From the beginning of the conversion process value assessment of the land will thus become a central concern of the *ejido*. Value assessment will be even more important to the *ejido* in the later stages, when it decides to commercialize or develop its property.

As soon as privatization is carried out, the *ejido* will be legally capable of marketing its developable and undevelopable land, within the restrictions that apply to each type. This stage poses important problems to the *ejido*. As poor agricultural communities which lack experience as urban development entrepreneurs, the *ejidos* will find themselves at a disadvantage. They will have to go through a learning process. Their lack of capital and capacity to hold out is another problem that the privatized *ejidos* will face. On the other side of the ledger, they will possess land of great economic value.

Various possibilities arise at this point. The *ejido* might decide to sell part or even all of its land in its undeveloped condition after taking care of the residential needs of its members. The land could be put up for sale through private business channels or through a public auction under the supervision of the government's central bank, with the right to refuse the highest bid. Another alternative would be that of linking itself to a partner in the private sector or to a government organization which would provide capital and know-how to develop the land, sharing in the risks and benefits of the subdivision venture. A combined strategy of selling part of the land in order to obtain resources to develop the rest is a third possible option that an *ejido* might find best.

To arrive at sound decisions, the *ejido* will need the preliminary advice of experts in land development and real estate marketing. While the *ejido* could seek such advice by contracting for the services of professionals, which might often be the best solution, the federal government should make available trustworthy and competent technical assistance to the *ejidos* in their difficult transition to the private sector. A central organization established by the suburban *ejidos* in the MAMC sharing common interests would be another possible source of technical assistance at this early stage.

As mentioned previously in the privatized *ejido* the *ejido* assembly will be transformed into a shareholders meeting where major decisions will have to be approved according to clearly stated procedural rules. The lack of internal democracy which is often attributed to the *ejido* poses a problem in this respect. If allowed to reproduce itself in the privatized *ejidos*, it can seriously impair the legitimacy and effectiveness of the new organization. Even assuming a perfected form of internal democracy in the *ejido* assembly, managing conflict within the *ejido* corporation can be seen as a very important concern. Access to the judicial review system and arbitration panels would also be necessary in the later stages. Partition of the land among dissenting groups, within the limits and restrictions of the planning system, should be available as a way out of internal conflicts in the *ejidos*.

The relationship that is established between the *ejidos* and the urban planning system would be pivotal in guiding the conversion and urbanization process of the *ejido*. Government, being vitally interested in implementing the reserve, will have to establish channels of communication to each privatized *ejido* to convey the short-and-long-term planning objectives and the operation of the planning system as it affects the *ejido*. This, however, would be meaningless if the traditional pattern of manipulation and powerlessness of the *ejido* vis-a-vis the different government organizations is not brought to an end in the urban peripheries. When the *ejido* finds that the channels of legality are accessible and rewarding, and that needless obstacles will not be placed in the way of its legitimate subdivision activity, it could be hoped that the *ejido* will internalize the planning objectives and act in accordance with them.

At the higher scale of a merger of various *ejidos*, one could expect to encounter some additional situations and problems, besides those previously described, which will now be briefly surveyed. When various *ejidos* are merged to create a privatized land holding corporation within the framework of the reserve declaration, a necessary preliminary step would be to assess the monetary value of the participating *ejido* to determine their proportional share in the total equity. One could expect that some of the participating *ejidos* would have a higher value than others due to their greater extension or superior characteristics. An unequal share distribution between the different *ejidos* would result, although within each *ejido* the internal distribution of the shares among its members would be equal. Profits would be distributed among the *ejidos* according to the shares held by each of them.

The voting rights held by the participating *ejidos* will reflect the shares held by them in the total equity. This would create a different situation than in the individual *ejido* corporations, where voting rights would be uniformly distributed among its members. Within the general assembly ruling the land holding corporation into which the separate *ejido* assemblies would become fused, voting rights would vary as a function of the shares held by each *ejido*. Only land not included in the reserve declaration and the privatization decree, if any, would maintain the existence of the previous *ejido* assemblies as independent entities.

The merger of a group of *ejidos* into a privatized land holding corporation would transfer the possession rights of the individual *ejidatarios* and *ejidos* to that body. Share holding property rights would become clearly separated from tenant arrangements. As long as no sales or development were to take place under the new organization, there would be no particular need for rearrangement of previous boundaries and activities or resettlement of *ejidatarios*. However, as soon as land begins to be sold, leased or developed, *ejidatario* tenant rights for agricultural and residential purposes would have to be renegotiated and rearranged internally.

Land classified within the ecological reserve that would continue in agricultural production permanently could be leased to those *ejidatarios* interested in continuing their rural life style. The same would also apply in the interim to developable land until its sale or development. Income from these leasings would obviously be distributed by the corporation to the shareholders. The rearrangements would also have to include establishment of a special residential area for the *ejidatarios*. As already mentioned the landholdings of the consolidated *ejido* corporation could be subdivided according to value among shareholder groups, if needed, within the limitations of the reserve and zoning statutes.

The Reserve and Prospective Growth of the MAMC

The global space needs for expansion in the MAMC that are projected for the next twenty years, and the region's potential to respond to them is an important question to be considered. A related question concerns the amount of *ejido* land that would have to be freed for this purpose. A study carried out under the coordination of Enrique Cervantes for the Comisión de Conurbación del Centro del País prior to its liquidation in 1987 (CCCP, 1986) has dealt with these questions, to which reference will be made in the ensuing discussion.[14]

The conurbation planning region of Mexico City was officially defined by SPP in a 1983 decree as comprising an area of 7,112.8 sq. km. in the following manner:

DF	16 delegations	1489 sq. km.	20.9%
SM	53 municipalities	5531 sq. km.	77.8%
Hidalgo	1 municipality	92 sq. km.	1.3%
Total	70 divisions	7112 sq. km.	100.0%

The CCCP study based its broad estimates of the total amount of land needed in the next two decades for MAMC expansion on the official population projections previous to the 1985 earthquake. According to these projections which were the accepted standard reference until the publication of the 1990 census results, the expected population of the MAMC for the years 2000 and 2010 were 24,935,000 and 32,063,000 people. Using the present average density as a standard, the CCCP study reached the conclusion that a new area of 453 sq. km. would have to be provided for urban expansion for the decade 1990-2000 and another 410 sq. km. for the decade following (CCCP, 1986: p. 64).

As already discussed in chapter 2, these figures are now considered outdated due to the important changes in the MAMC population trends during the past decade. On the assumption that the 1980-1990 trend will continue during this decade and the next, the population and area growth figures can be expected to be considerably smaller—a possibility which will make it easier to manage growth in the MAMC. Refering again to chapter 2 (see tables 3 and 4) population growth under such an assumption would add 1,574,000 people during this decade and 1,734,000 during the next. The areas required would be 205 sq. km. for the 1900-2000 decade and 224 sq. km. for the next.

The next question to consider concerns the availability of land within the conurbation region to provide for these needs. The *Declaratoria de Usos y Destinos* issued by the DF government in 1985 decreed the following distribution of land uses within its borders:[15]

Urban Area	554.0 sq. km.	37.2%
Urban Growth Reserve	79.8 sq. km.	5.4%
Ecological Reserve	855.5 sq. km.	57.4%
Total DF Area	1489.3 sq. km.	100.0%

The lands designated as an ecological reserve play a vital

environmental role particularly as an aquifer recharge area. However, the 1990 inventory of DF land shows that there are only 808 sq. km. of rural land left within the DF in 1990, 54.2 % of its total area, which clearly indicates that not only has the growth reserve been completely used up, but that the ecological reserve has already suffered a very substantial loss. While urban expansion within the DF boundaries has diminished, it nevertheless provided 25% of total expansion of the MAMC in the last decade, reducing these ecologically vital lands. The limit of 633.8 sq. km. of urbanized land adopted by the DF government has already been surpassed in the DF which has reached an extension of 681.3 sq. km. in 1990 (table 2).

Table 18. Urban Land Reserves (sq. km.) Designated by the SM Government for MAMC Growth (1985)

Locality	Reserve Area
STRATEGIC CENTERS	
1) Cuautitlán Izcalli	12.9
2) Chalco	5.1
3) Ecatepec	26.5
4) Netzahualcóyotl	0.9
PERIPHERAL CITIES:	
1) Atizapán	23.4
2) Coacalco	9.6
3) Cuautitlán	3.3
4) Chimalhuacán	6.7
5) Chicoloapan	3.9
6) Ixtapaluca	6.4
7) La Paz	6.8
8) Nicolás Romero	8.1
9) Tecamac	18.3
10) Tultitlán	11.5
11) Amecameca	9.8
12) Huehuetoca	13.4
13) S. M. de las Piramides	4.3
14) San Juán Teotihuacán	8.4
15) Tepotzotlán	2.1
16) Texcoco	7.7
17) Zumpango	11.3
TOTAL	200.0

Source: Programa de Reservas Territoriales de la ZMCM, CCCP (1985)

The SM government on its part has allocated 200 sq. km. as an urban growth reserve to accommodate the expansion needs of the MAMC (table 18). The reserve area which is distributed in four strategic centers and seventeen peripheral towns would increase the 1990 SM metropolitan area to 824.1 sq. km. which is the growth objective adopted by the SM government for its conurbated area (table 19). This area could accomodate the expansion needs until the end of the 1990-20000 decade but would hardly go beyond that.

Table 19. Land Use Objectives for the Conurbation Region of the MAMC Proposed by the DF and SM Governments

Land Use	Total Conurbation (sq. km.)	(%)	DF (sq. km.)	SM (sq. km.)	Hidalgo (sq. km.)
Urbanized Areas	1466.3	20.6	633.8	822.0	10.5
Texcoco Lake	177.0	2.5	0.0	177.0	0.0
Forests	1887.9	26.5	635.0	1252.9	0.0
Seasonal Agriculture	2402.7	33.8	203.5	2130.0	69.2
Irrigation Agriculture	424.7	6.0	16.0	408.7	0.0
Natural Areas	741.4	10.4	1.0	740.4	0.0
Other	12.8	0.2	0.0	0.0	12.8
Total Area	7112.8	100.0	1489.3	5531.0	92.5

Source: Proyecto del Programa Rector de Usos del Suelo para la ZMCM, DF and SM Governments, 1985 as q. CCCP,1986: p. 10, table II.

The CCCP study explored the availability of additional land in the conurbation region to supply the reserve beyond the previous growth objectives. The study adopted natural aptitude criteria based on slopes, disaster risk, ecological value and agricultural productivity to identify this land. The aptitude criteria were also combined with availability of regional infrastructure to establish a priority order in the incorporation of the land to the urban reserve (CCCP, 1986: p. 34-37).

The slope criterion excluded from the growth reserve estimate any land with a gradient above fifteen percent due to its higher development cost and higher land-related disaster risks. The ecological conservation criterion eliminated the six national and state parks and some smaller ecological reserves in the region (651 sq. km.) as well as the ecological reserve in the south of the DF (855 sq. km.) previously mentioned. The risk criterion led to the exclusion of land with high seismic risk or periodic flooding. Finally, the productivity criterion led to the exclusion of land with higher agricultural productivity (CCCP, 1986: p. 34-37).

The land areas selected as appropriate for the reserve on the basis of these natural criteria are located within 29 of the 54 municipalities of the conurbation planning region. They yielded an estimate of another 685 sq. km. which together with the reserve area previously mentioned would add up to a total urbanized area of 2290 sq. km., roughly 36.7% of the total extension of the SPP conurbation planning region (table 19). Assuming the continuation of the present population trend such an area should accomodate growth for another 30 years within the next century.

The further application of the infrastructure criteria to detect the land with the highest infrastructure endowment identified ten of these municipalities, highlighted in bold letters, as the most appropriate for the immediate expansion needs of the MAMC in the decade to come (table 20). The CCCP study found that the share of *ejido* land in the total conurbation planning region and within the CCCP reserve area is 51.9% (tables 20 and 21). That percentage is remarkably close to the one found for the MAMC growth area by such authors as Schteingart and Iracheta. These figures can convey an approximate idea of the amount of *ejido* land that on the average would have to be privatized periodically and incorporated into the reserve to respond to expansion pressures.

The conclusion to be reached from the CCCP estimate is that an adequate reserve of land is available in the conurbation region to accommodate expansion of the MAMC for various decades to come. This could be achieved with full regard for the principles of environmental planning. Although the CCCP study does not come out with an explicit formal vision of future growth of the MAMC beyond estimating potential reserve land, it certainly allows for a different form of growth than that of urban sprawl, as in the past. A pattern of growth based on nucleated deconcentration, whereby a great number of autonomous smaller towns within the conurbated region would absorb future demographic and economic growth is a real possibility. A local balance of people and jobs in these towns would be one of the main features of such a scheme, without, however, unrealistically excluding commuter movements. Such movement would require major expansion of the present limited regional communications and transportation network.

While concerns for the environmental impact of these towns on the now-isolated rural areas of the conurbation region are justified, such concerns should not be allowed to distort the issues involved. There is much to be said in favor of a scheme of nucleated deconcentration, both from the standpoint of the central city and the new towns. The metropoli has already reached the breaking point in the concentration of industries, vehicles, and people. This has turned the MAMC into one of the most contaminated and congested areas in the world.

Table 20. Areas Qualified for Urban Development in 28 of the Municipalities of the Conurbation, by Municipality and Tenure Status (sq. km.)

Jurisdiction	Total Area of Municipality	Developable Land Area	Developable Land (%)	Tenure of Land *Ejido*	Tenure of Land Non-*Ejido*
1) Acolman	95.4	14.8	(15.5)	12.7	2.1
2) **Amecameca**	164.0	39.1	(23.8)	8.5	30.7
3) Atenco	94.3				
4) Atizapan	79.8				
5) Atlautla	155.3	15.8	(10.2)	13.1	2.7
6) Axapusco	274.8	82.6	(30.0)	59.9	22.8
7) Ayapango	34.7	27.4	(79.0)	14.3	13.1
8) Coacalco	40.4				
9) Cocotitlan	10.5	8.2	(78.1)	0.6	7.6
10) Coyotepec	12.3				
11) Cuautitlan	36.0				
12) C. Izcalli	105.5				
13) **Chalco**	284.6	49.6	(17.4)	25.6	23.9
14) Chiautla	19.7	2.8	(14.2)	1.7	1.1
15) **Chicoloapan**	61.2	9.5	(15.5)	5.8	3.7
16) Chiconcoac	7.2				
17) Chimalhuacan	44.7				
18) Ecatepec	155.1				
19) Ecatzingo	54.1				
20) **Huehuetoca**	158.3	18.7	(11.8)	7.9	10.8
21) Huixquilucan	108.2				
22) I. Fabela	80.4				
23) **Ixtapaluca**	334.7	1.7	(0.5)	0.6	1.1
24) Jaltenco	3.8				
25) Jilotzingo	120.2				
26) Juchitepec	151.7	15.6	(10.3)	10.5	5.1
27) Ocampo	17.2				
28) Naucalpan	123.2				
29) Netzahualcóyotl	63.2				
30) Nextlalpan	69.6				
31) N. Romero	214.4				
32) Nopaltepec	87.1	49.6	(56.9)	36.4	13.2

33) Otumba	140.8	11.1	(7.9)	0.5	10.6
34) Ozumba	47.7	35.8	(75.0)	11.1	24.7
35) Papalotla	3.3	2.1	(63.6)	0.0	2.1
36) La Paz	34.4				
37) **SM Piramides**	73.3	8.2	(11.2)	4.7	3.5
38) **Tecamac**	154.9	5.6	(3.6)	0.0	5.6
39) Temamatla	29.1	12.5	(42.9)	3.1	9.4
40) Temascalapa	177.3	76.6	(43.2)	33.6	43.0
41) Tenango	41.8	14.6	(34.9)	7.2	7.4
42) Teoloyucan	33.3				
43) **Teotihuacan**	93.6	20.6	(22.0)	6.7	14.0
44) Tepetlaoxtoc	157.4				
45) Tepetuxpa	53.7	19.8	(36.9)	8.2	11.6
46) Tepotzotlan	222.8				
47) **Texcoco**	411.0	20.6	(5.0)	7.6	13.0
48) Tezoyuca	11.1	1.2	(10.8)	0.0	1.2
49) Tizayuca	64.4	58.0	(90.1)	58.0	0.0
50) Tlalmanalco	129.4	22.8	(17.6)	10.6	12.2
51) Tlalnepantla	87.7				
52) Tultepec	37.2				
53) Tultitlan	56.2				
54) **Zumpango**	227.8	31.6	(13.9)	5.2	26.4
Total	5549.2	685.2	(12.3)	355.8	329.4

Source: Programa de Reservas Territoriales de la ZMCM, CCCP (1985)

Table 21. Type of Land Tenure in the MAMC Growth Area (1940-1982), CCCP Planning Region (1985) and in the Proposed Reserve (sq. km.)

Type of Tenure	MAMC Growth«		CCCP Regionß		CCCP Reserveµ	
	Area	%	Area	%	Area	%
Ejidal and Comunal	470.9	51.5	2880.6	51.9	355.8	51.9
Private	349.0	33.8	2312.6	41.7	326.1	47.6
Federal and Unidentified	93.9	9.1	356.0	6.4	3.2	0.5
Total	913.8	100.0	5549.2	100.0	685.1	100.0

Sources: (CCCP,1985: p. 38). « DF + 11 SM municipalities (1940-1975). ß 54 SM municipalities and Hidalgo, excludes the DF. µ Includes the reserve proposed by the CCCP, without the 279.8 sq. km. proposed by the DF and EM governments.

Nucleated deconcentration could slow further deterioration. The human scale and manageable size of the smaller towns, their contact with nature, and their deconcentration of contamination sources constitute clear advantages over suburban sprawl. While one could hardly object to the conservation of agricultural land as being a worthy objective, it may often be necessary to sacrifice agricultural land for urban growth where there are pressing urban needs to be satisfied and the more vital objectives of disaster prevention and ecological protection are at stake.

Although linking *ejido* land conversion to the reserve concept, as envisioned in the proposal, is an attractive proposition from the planning standpoint, the two concepts can be treated independently, if necessary. Thus, even if the implementation of an urban reserve were to be left pending, expropriation of the *ejidos* in the terms suggested in the proposal can still make a positive contribution to substantially reduce illegal conversion and irregular development of *ejido* land. Expropriation of the land in favor of the *ejidatarios* and settlement Trust funds as need for urban expansion arises would still constitute a far superior solution over an urban growth policy based on confiscatory *ejido* expropriation, and a housing policy based on chaotic illegal settlement and regularization.

The expropriation procedure should, in such a situation, contain by itself the necessary safeguards for a selective process of land aptitude to operate. It will have to be proven that the land to be privatized and developed has no fundamental ecologic or safety impediments for urban use wherever *ejido* expropriation is carried out on grounds of urban development needs. A more explicit clause to Article 112 of the Federal Agrarian Law requiring that proof be provided that any *ejido* land expropriated on grounds of urban development needs has no fundamental ecologic or safety impediments for urban use would be required. The expropriation procedure itself, independent of the reserve, would then be playing a most valuable planning function that would preclude from expropriation and conversion any type of land that lacks minimal environmental and safety requirements.

CONCLUSIONS

The results of this study can be seen to have a significance going beyond the particular problem of the *ejidos* in the urban peripheries. It should help to recognize the crucial importance of secure property rights in economic development, particularly for the Third World countries. The evidence gathered leads to the clear conclusion that the highly restricted property rights of the *ejidos* have stood in the way of achieving a more efficient land conversion process in the peripheries of Mexican cities. The non-marketability and the highly inequitable terms on which the expropriation of these property rights takes place have brought about patterns of behavior on the part of the *ejidos* and of government, which have been highly detrimental to the urbanization process.

The urbanization process has largely escaped the framework of legality and the planning system, being transferred to the realm of the informal sector, with critical effects on the city's environment and quality of life. Much of the land of the *ejidos*, often highly inadequate for urban development purposes, is being illegally sold and chaotically occupied by irregular settlements. The overwhelming majority of the irregular settlements have established themselves on illegally converted *ejido* land, while the percentage of *ejido* land occupied by irregular settlements is the highest among all types of property. Irreplaceable land of unique ecological value is being lost. Land unfit for urban development is being occupied. The non-marketability of *ejido* land, and the near confiscatory terms of *ejido* expropriation by government, have contributed much to this unfortunate outcome.

Irregular settlement activity has been encouraged by the state to consolidate a major consumer and labor market in the metropolitan region as the basis of national industrialization, at a minimal immediate direct cost to the treasury. The settlements have provided the official political establishment with a captive political clientele dependent on government favors. Seen from the perspective of the urban settlers, informal settlements proved to be the most striking informal survival strategies of the low income people in the urban environment. They provide living proof of the great potential that lies in people's own initiatives and labor outside of the formal economy.

Although this housing formula may initially seem to be without cost, it has very high costs for the settlers themselves as well as for society as a whole—which are directly related to their lack of recognized property rights to the land. Economic value of the properties in the informal settlements is reduced to a fraction of comparable properties in the formal market, investment is greatly curtailed and delayed, economic mobility of the land and buildings is hindered and formal credit cannot be obtained. The lack of a tax base in the settlements undermines government capacity to improve their services. Settlers must concentrate on political activity to win recognition of their property rights instead of using their energy on productive activities to improve their income and settlement.

A vicious circle, based on the lack of property rights, perpetuates itself from the *ejidos* to the irregular settlements in informal society. The people in the informal settlements are forced to begin at the wrong end. In the reverse order of what occurs in formal development irregular settlers occupy the land, then build their dwellings, much later introduce minimal services and infrastructure, and only at the end do they finally acquire title to their properties through regularization. There can hardly be a stronger indictment against a legal system which has created a situation in which 65% of its population has been forced to resort to this type of reverse order development, as is the case in the MAMC.

Full property rights for both *ejidos* and settlers are necessary from the earliest beginning of urbanization to break this vicious circle and bring about a more efficient land conversion process in the urban peripheries. *Ejidatarios* and settlers should be entitled to benefit from the enhanced value and opportunities of possessing property rights to the land, while society should be in a position of regaining land use control over the land conversion process and economic value for tax purposes. Recognized fee simple property rights to *ejidos* and settlers from the very beginning of the urbanization process can turn the new situation into a positive sum game where almost everybody stands to gain over time.

Authors such as Ronald Coase, have contributed much to highlight the role of property rights in economic development. They have shown that undefined property rights raise the transaction costs, inhibit capitalization and obstruct economic and social progress. One can refer as well to Douglas C. North (1986) who has argued convincingly on the major historical role that secure property rights and freedom played in modern economic growth in the Western developed nations and how stagnation set in where these conditions were lacking.[16]

Awareness on the role of property rights is also beginning to gain ground in the Third World development literature, specifically in relation to the informal economy. As greater understanding of the informal sector's great significance for the economy and political stability of the Third World countries is acquired, it is becoming increasingly clear that informality has very high costs as a result of the lack of secure property rights guaranteeing the inviolability of people's investments. As the value of their assets is greatly diminished in the informal economy, people do not develop their resources as efficiently as they might, and their incentives to invest are reduced. Their efforts and initiatives are curtailed and diverted from productive activities and wasted instead on overcoming the obstacles that stand in the way of the formal recognition of their property rights.[17]

The Third World nations face the challenge of incorporating their informal economies within the legal framework without destroying them. The vital task of adjusting their legal systems to allow people to pursue their legitimate activities within the law still lies ahead of them. If these societies are to progress they will have to create an environment, where secure property rights and freedom is guaranteed to all groups and individuals. This is a precondition for economic development.

The quest for greater political and economic freedom, should not, however, obscure the need for government regulation and government assistance were necessary. There are great limitations to what government and people can achieve in isolation and great potential for what they can achieve in concert. Regulation becomes necessary where the market produces results which are clearly against the recognized general interests of society; government assistance is required, on the other hand, where the market ceases to provide for people's elementary needs. Government responsibilities and people's initiatives must cease to be seen as contradictory terms by an enlightened society. They should complement each other. The great challenge to third World societies is how to maintain state intervention within its legitimate minimum boundaries while maximizing freedom and opportunities for people to carry out their initiatives.

In pursuing a solution to the chaotic situation in the peripheries of Mexican cities, this proposal seeks such an equilibrium. Government is assigned the vital role of regulating the process of incorporation of *ejido* and private land to the cities and administering the transfer and habilitation of part of that land to house low-income people. At the same time, undiminished property rights and freedom to pursue their best interests are provided within this framework to *ejidatarios*, private owners and settlers. The necessary conditions are being created to induce people's own initiatives in the development of their habitat.

The search for a formula which recognizes the legitimate role of government without diminishing people's rights and freedoms is meaningful for Third World societies in their difficult journey to economic development. A formula where people, all people, can live, work and prosper in political and economic freedom within the framework of the law in every area of their lives, and not at its margin, is certainly one of the worthy goals that the developing world can choose to pursue.

---

[1] Even if compensation as a function of agricultural value would still have to be paid to the *ejidos* in terms of the present legislation, the proposal carries great significance.

[2] (1983: p. 34).

[3] (Turner, 1963; 1966; 1972; 1976).

[4] See tables 6 and 7 of ch. II. Iracheta, 1984: table 5; p. 59.

[5] According to José Merino Castrejón, Director of Land Regularization in the DF government there are approximately 500,000 plots in the DF alone which are still not

legalized, and not incorporated into the city's tax base (Excelsior, June 14 and 15, 1990)

[6] Assessment of real property values would have to reflect the commercial value of the land and the actual level of infrastructure and services that characterize the settlement at each stage.

[7] (Guerrero et al. 1974; de la Rosa, 1974; Schteingart, 1978; 1982; 1983; Iracheta, 1984).

[8] (Gilbert & Ward, 1985: p. 250-251).

[9] (Bataillon, 1968: p. 25; Turner & Fichter, 1972: ch. 7).

[10] (Violich, 1981).

[11] In an ordered planning process these conservation and vulnerable sites would be identified by officially sanctioned physiographic surveys. These surveys should be made mandatory in the future by the Human Settlements Law.

[12] The literature refers to such models in terms of their geometric form, such as linear, star, concentric, nucleated, etc. The growth model adopted by the Plan Nacional de Desarrollo Urbano in 1983 (PND, 1983) for MAMC is one which seeks deconcentration. This form contemplates a contained central core surrounded by a metropolitan system of smaller nucleated urban centers placed within well-protected natural surroundings.

[13] (Mc' Harg, 1969).

[14] (CCCP, 1987).

[15] (CCCP, 1986: p. 49).

[16] (North, 1986).

[17] (de Soto, The Other Path, 1989).

BIBLIOGRAPHY

Adler, Judith Hellman. 1983. *Mexico in Crisis.* New York: Holmes and Meier Publishers.

Adler, Larissa Lomnitz. 1977. *Networks and Marginality: Life in a Mexican Shantytown.* New York: Academic Press.

Aguilar Camín, Hector and Meyer, Lorenzo. 1990. *A la Sombra de la Revolución Mexicana.* México, DF: Aguilar, León y Cal Editores SA de CV.

Aguirre Avellaneda, Jerjes. 1976. *La Politica Ejidal en Mexico.* Mexico, DF: Instituto Mexicano de Sociologia.

Angel, Shlomo et al. 1983. *Land for Housing the Poor.* Singapore: Select Books.

Armnstrong, Warwick and McGee, T. G. 1985. *Theatres of Accumulation, Studies in Asian and Latin American Urbanization.* London and NY: Methuen.

Azuela de la Cueva, Antonio. 1989. *La Ciudad, La Propiedad Privada y el Derecho.* México DF, El Colegio de México.

Bataillon, Claude. 1968. *Las Zonas Suburbanas de la Ciudad de México.* México, DF: Instituto de Geografía, UNAM.

Bennett, Mark. 1986. *Public Policy and Industrial Development.* Boulder, CO and London: Westview Press.

Blair, Thomas. 1983. "Third World Urban Innovation: Managing Growth and Development." *Cities.* August 1983): 71-84.

Blumenfeld, Hans. 1967. *The Modern Metropolis: Its Origins, Growth Characteristics and Planning.* Cambridge, MA: MIT Press.

Bourne, L.S. 1984. *Urbanization and Settlements Systems: International Perspectives.* New York: Oxford University Press.

Breeze, G. 1985. *The City in Newly Developing Countries: Readings on Urbanization.* Engelwood Cliffs NJ: Prentice Hall.

Brennan, E.M. 1985. "Contradictions in Mexico's National Policy." *Regional Development Dialogue 4*, 1982: p. 21-24

Brotchie, John; Newton, Peter; Hall, Peter; and Nijkamp, Peter, Ed. 1985. *The Future of Urban Form.* London: Croom Helm.

Brower, J. David. 1976. *Urban Growth Management Through Development Timing*. New York: Praeger.

Brown, Jane Cowan. 1972. *Patterns of Intra-Urban Settlements in Mexico-City: An Examination of the Turner Theory*. Ithaca, NY: Cornell University Press, Latin American Studies Program Dissertation Series, 40.

Cabrera, A. Gustavo. 1982. "Politica Demografica sobre Migración Interna." *Revista Demografia y Economia*. XVI:3. México, DF: (1982): 439-448.

Campbell, Tim, and Wilk, David. 1986. "Plans and Plan Making in the Valley of Mexico: The Evolution of a Planning Process, 1970-1985." *Third World Planning Review*. 8:(4) (1986): 287-313.

Cárdenas, Cuauhtémoc. 1973. "Algunas Ideas para una Reforma Urbana en México." *Sociedad Mexicana de Panificación*. México, DF, Julio, 1973.

Carrillo, Azpeitia Rafael. 1984. *Historia de la Ciudad de Mexico: Desde su Fundacion como Capital del Reino Azteca hasta su Gran Desarrollo Actual.*. México, DF: Panorama Editorial.

Castells, Manuel. 1979. *Multinational Capital, National States, and Local Communities: A Report on Urbanization and Social Change in the Third World based on Field Work Research in Squatter Settlements.* Berkeley: (IURD, UC).

Castells, Manuel. 1981. *Crisis Urbana y Cambio Social.* Mexico, DF: Siglo XXI.

CCCP, 1986. *Programa de Reservas Territoriales de la ZMCM*. México, DF: Mimeo.

Chevalier, Francois. 1963. *Land and Society in Colonial Mexico: The Great Hacienda*. Berkely: University of California Press.

CIHAC, 1988. *Catalogo 88 de la Construcción*. Ed. Eduardo Carrera Cortés. México, D.F.

CIHAC, 1991. *Catalogo 88 de la Construcción*. Ed. Eduardo Carrera Cortés. México, D.F.

Coleman, S. James. 1974. *Power and Structure of Society*. New York: W.W. Norton & Co.

CONAPO. 1984. *(Evolucion y Perspectivas Demograficas de la Zona Metropolitana de la Cd. de Mexico)*. México, DF.

Connolly, Priscilla. 1976. *Analisis del Comportamiento del Mercado de Bienes Raíces en la Zona Metropolitana de la Cd. de México*. COPEVI, México, DF.

Connolly, Priscilla. 1981. *Towards an Analysis of Mexico City's Local State* (Preliminary Notes for Discussion), México: mimeo.

COPEVI, Centro Operacional de Vivienda y Poblamiento AC. 1977. *La Producción de Vivienda en la Zona Metropolitana de la Ciudad de Mexico*. Mexico, DF: COPEVI.

COPEVI, Centro Operacional de Vivienda y Poblamiento AC. 1977. *Las Politicas Habitacionales del Estado Mexicano, Investigaciones sobre Vivienda*. Mexico, DF.

Cornelius, A. Wayne, and Kemper, Robert. 1978. *Metropolitan Latin America: The Challenge and the Response*, Latin American Urban Research, Vol.6. Sage Publications.

Cornelius, A. Wayne. 1975. *Politics and the Migrant Poor in Mexico City*. Stanford, CA: Stanford University Press.

Cornelius, A. Wayne. 1976. "The Impacts of Cityward Migration on Urban Land and Housing Markets." *The City in Comparative Perspective* , eds. John Walton & Louis Masotti. Beverly Hills, CA: Sage Publications.

Cruz Rodriguez, Ma. Soledad. 1981. "El *Ejido*, en la Urbanización de la Ciudad de México." Tésis Profesional, UAM Azcapotzalco, México, DF. mimeo.

Cumberland C. Charles. 1972. *Mexican Revolution: The Constitutionalist Years*. Austin & London. University of Texas Press.

Davis, E. Lance, and North, C. Douglass. 1971. *Institutional Change and*

American Economic Growth. New York: Cambridge University Press.

De la Rosa, Martín. 1974. *Netzahualcóyotl, un Fenómeno*. México, DF: Testimonios, Fondo de Cultura Económica.

De Soto, Hernando. 1989. *The Other Path: The Invisible Revolution in the Third World*. New York: Harper & Row.

Demsetz, Harold. 1967. "Toward a Theory of Property Rights." *American Economic Review*. (May, 1967).

Diario de los Debates del Congreso Constituyente, 1916-1917. 1917 re. 1960. México DF: Talleres Gráficos de la Nación.

Díaz Ramírez, Manuel. *"Apuntes del Movimiento Obrero y Campesino de México, 1844-1880"*. México ed. FCP, 1938. (q. Aguirre Avellaneda, 1976).

Drakakis-Smith, David. 1986. *Urbanization in the Developing World*. London: Croom Helm.

Durán, M.A. 1967. *El Agrarismo Mexicano*. México, DF: Siglo XXI.

Durand, Jorge. 1983. *La Ciudad Invade el Ejido* . México, DF: ediciones de la Casa Chata, Cultura SEP.

Dwyer, D.J. 1974. *The City in the Third World*. New York: Barnes and Noble.

Eckstein, Salomón. 1966. *El Ejido Colectivo en México*. Mexico, DF: Fondo de Cultura Económica.

Eckstein, Susan. 1977. *The Poverty of Revolution: The State and the Urban Poor in Mexico*. Princeton, NJ: Princeton University Press.

Fernandez y Fernandez, Ramón; Acosta, Ricardo. 1961. *Politica Agrícola: Ensayo sobre Normas para México*. Mexico DF: Fondo de Cultura Económica.

Frieden, B. 1965/66. "The Search for a Housing Policy in Mexico City." *Town Planning Review*, Vol. 36 (1965/66): 75-90

Friedman, Milton, and Friedman, Rose. 1980. *Free to Choose*. New York: Harcourt, Brace, Jovanovich.

Furtak, Roberto K. 1974. *El Partido de la Revolución y la Estabilidad Politica en México*. México: UNAM, CPS serie estudios 35.

Furuboton, G. Eirik, and Pejovich, Svetozar. 1972. "Property Rights and Economic Theory: A Survey of Recent Literature." *Journal of Economic Literature:* 1137-1162.

Garcia Peralta, Beatriz. "Estado y Capital Privado en el Fraccionamiento Izcalli-Chamapa." 1983. *Revista Mexicana de Sociologia*. 43,4. (Oct-Dec. 1983):1439-1464.

Garza, Gustavo and Schteingart, Martha. 1978. "La Acción Habitacional del Estado en México," El Colegio de México, 1978.

Garza, Gustavo. 1984. "Concentracion Espacial de la Industria en la Cd. de Mexico. 1930-70." *Revista Demografia y Economia*. XVIII (1-1984): 3-26 .

Garza, Gustavo. 1981. "La Problematica de la Ciudad de México," en Centro de Estudios Económicos y Sociales del Tercer Mundo. *Lecturas del CEESTEM*. Vol. 1-3, 1981.

Garza, Gustavo. 1982. "México: Politica Demográfica Sobre Migración Interna (1978-1982)" *Revista Demografia y Economia*. XVI (3-1982): 439-448

Gentelman, Judith. 1984. *Mexican Oil and Dependent Development*. New York: Peter Lang Publishing Inc.

Gilbert, Alan and Gugler, Josef. 1982. *Cities, Poverty, and Development: Urbanization in the Third World.* New York: Oxford University Press.

Gilbert, Alan and Ward, Peter. 1985. *Housing, the State and the Poor: Policy and Practice in Three Latin American Cities*. Cambridge, England: Cambridge University Press.

Gilbert, Alan. 1982. *Urbanization in Contemporary Latin American. Critical Approaches to the Analysis of Urban Issues*. New York, John Wiley and Sons.

Goldscheider, Calvin. 1983. *Urban Migrants in Developing Nations*. Boulder, CO: Westview Press.

Gonzalez Casanova, Pablo y Florescano Enrique. 1979. *México Hoy*. México DF: Siglo XXI.

Gonzalez Salazar, Gloria. 1983 *El Distrito Federal: Algunos Problemas y su Planeacion*. Mexico, DF: UNAM.

González Navarro, Moisés. 1968. *La Confederación Nacional Campesina*. México, DF: B. Costa-Amic.

González Navarro, Moisés. 1974. *México, Población y Sociedad (1900-1970)*, México, DF: UNAM.

Grindle, S. Merilee. 1977. *Bureaucrats, Politicians, and Peasants in Mexico: A Case Study in Public Policy*. Berkely, CA: The University of California Press.

Grindle, S. Merilee. 1986. *State and Countryside. Development Policy and Agrarian Politics in latin America*. Baltimore, MD: The Johns Hopkins Press.

Guerrero, Maria Teresa et al. 1974. *La Tierra, Especulacion y Fraude en el Fraccionamiento Nuevo Paseo de San Agustin*. Mexico, DF: Mimeo.

Guerrit, Huizer. 1968. *Los Movimientos Campesinos en México*. Mexico DF:

Gutelman Michel. 1971.*Capitalismo y Reforma Agraria en México*. México DF: Editorial Era.

Hamilton, Nora and Harding, Timothy F. 1986. editors. *Modern Mexico: State, Economy, and Social Conflict*. Beverly Hills, CA: Sage Publications.

Hansen, R. 1971. *The Politics of Mexican Development*. Baltimore, MD: The Johns Hopkins Press.

Hauser, M. Philip. 1982. *Population and the Urban Future*. Albany, NY: The State University of New York Press.

Hirschmann, Albert O. 1984. *Getting Ahead Collectivelly, Grassroots Experience in Latin America*. New York: Pergamon Press.

Hirschmann, Albert O. 1986. *Rival Views of Market Society and other Recent Essays*. New York: Elisabeth Sifton Books. Viking Penguin Inc.

Hodges, Donald and Gandy, Ross. 1983. *Mexico 1910-1982: Reform or Revolution?*. London: Zed Press.

INEGI. 1984. *Anuario Estadistico del DF 1984*. México DF: INEGI.

INEGI. 1991. *Area Metropolitana de la Cd. de México (AMCM). Sintesis de Resultados.* XI Censo General de Población y Vivienda. México.

INEGI. 1992. *Ciudad de México (area Metropolitana): Resultados Definitivos Tabulados Basicos.* XI Censo General de Población y Vivienda. México.

Iracheta, X. Alfonso. 1984. *El Suelo, Recurso Estrategico para el Desarrollo Urbano.* Toluca: Universidad Autonoma del Estado de Mexico.

Jacobs, Jane. 1985. *Cities and the Wealth of Nations.* New York: Vintage Books.

Jury, Salvador. 1983. "Aciertos y Desaciertos en la Ciudad de México." *Revista Mexicana de Ciencias Politicas y Sociales.* (Jan-June, 1983): 183-200.

Katz, Friedrich. 1981. *The Secret War in Mexico: Europe, the US and the Mexican Revolution.* Chicago: The University of Chicago Press.

Kemper, V. Robert. 1977. *Migration and Adaptation. Tzintzuntzan Peasants in Mexico City.* Beverly Hills, CA: Sage Publications.

Large, W. Donald. 1973. "This Land is Whose Land? Changing Concepts of Land as Property." *Wisconsin Law Review,* (1973): 1039-1083.

Leff, Arthur Allen. 1974. "Economic Analysis of Law." *Virginia Law Review,* Vol. 60. (1974).

Linn, F. Johannes. 1983. *Cities in the Developing World: Policies for their Equitable and Efficient Growth.* New York: World Bank Publication, Oxford University Press.

Logan, R. John, and Molotch, L. Harvey. 1987. *Urban Fortunes.* Berkeley, CA: University of California Press.

Lomnitz Adler, Larissa. 1977. *Networks and Marginality: Life in a Mexican Shantytown.* New York: Academic Press.

López Rivera, Cecilia. 1978. *La Intervención del Estado en la Formación de un Asentamiento Proletario: El Caso de la Colonia Ajusco.* Tésis Profesional, UIA. México DF: mimeo.

Lugo Medina, Ma. Luisa, and Bejarano González, Fdo. 1981. *La Acción del Estado, el Capital y la Formación de las Colonias Populares, en la Transformación Urbana de las Tierras Ejidales en las Delegaciones de*

Magdalena Contreras y Tlalpan. El Caso de la Colonia Popular Miguél Hidalgo. Tésis Profesional, UIA. Mexico, DF: mimeo.

Mc'Harg, Ian. 1969. *Design with Nature*. Philadelphia, PA: Falcon Press.

McGee, T.G. 1979. *The Urbanization Process in the Third World: Explorations in Search of a Theory*. London: Bell and Hyman.

McKean, R. 1972. "Property Rights Within Government and Devices to Increase Government Efficiency." *Southern Economic Journal*, (October): 177-186.

Mendieta y Nuñez, Lucio. 1964. *El Problema Agrario de México*. México, DF: Editorial Porrua.

Michelman, I. Frank. 1981. "Property as a Constitutional Right." *Washington and Lee Law Review*, 1097 (1981).

Miller, M. Mark. 1988. *Managing the Maelstrom: Regional Planning Evaluation for the Mexico City Metropolis and the Mexican Urban System*. Paper presented to the Association of American Geographers. Phoenix, Arizona: mimeo.

Molina Enríquez, Andrés. 1909 re. 1978-1991. *Los Grandes Problemas Nacionales*, México, DF: Ediciones Era.

Molina Enríquez, Andrés. 1937. *La Revolución Agraria en México,* México, DF. unknown.

Mortera Ortega, Dora María. 1981. *La Propiedad de la Tierra en Santiago Ahuizotla, Historia de un Poblamiento: 1929-1974*. Tesina, Universidad Autonoma Metropolitana. México DF: mimeo.

Muñoz, Humberto; De Oliveira, Orlandina; Stern, Claudio. 1982. "Mexico City: Industrialization, Migration and the Labor Force, 1930-1970." Paris: *UNESCO, Selected Studies on the Dynamics, Patterns and Consequences of Migration*. (no.1, 1982).

Newell, R. and Rubio, L. 1984. *Mexico's Dilemma: The Political Origin of Economic Crisis*. Boulder, CO: Westview Press.

Niemeyer, E.V. Jr. 1974. *Revolution at Queretaro: The Mexican Constitutional Convention of 1916-1917*. Austin, TX: The University of Texas Press.

North, C. Douglass. 1986. Institutions, Economic Growth and Freedom: An Historical Introduction. *Political Economy Working Papers*. St. Louis, MO. Washington University.

Pacione, Michael, Ed. 1981. *Problems and Planning in the Third World Cities*. New York: St. Martins Press.

Paviani, Aldo. 1985. "La Urbanizacion en America Latina. El Proceso de Constitucion de Periferias en las Areas Metropolitanas." *Revista Interamericana de Planificacion*. Vol. 19-73, (Marzo 1985).

Perló, C. Manuél. 1979. "Politica y Vivienda en México 1910-1952." *Revista Mexicana de Sociologia..*

Perrello, Sergio. 1975. "San Miguel de la Labor, o las Sobras de una Hacienda." *Revista Mexicana de Sociologia*. 37,3 (Jul-Sept. 1975):717-736.

Pommier, Paulette. 1982. "The Place of Mexico City in the Nations Growth: Employment Trends and Policies." *International Labour Review*. 121 (May-June 1982):345-360.

Portes, Alejandro & Walton, John. 1976. *Urban Latin America*. Austin, TX: University of Texas Press.

Potter, B. Robert. 1985. *Urbanization and Planning in the Third World. Spatial Perceptions and Public Participation*. London: Croom Helm.

Puente Leyva, Jesús: 1975. "Sector Agricola: Tendencias Recientes y Perspectivas." Ponencia; Oaxtepec, Morelos. mimeo.

Rodgers, H. William Jr. 1982. "Bringing People Back. Toward a Comprehensive Theory of Taking in Natural Resources Law." *Ecology Law Quarterly* , 205: 385-432.

Rouaix, Pastor. 1959. *Génesis de los Articulos 27 y 123 de de la Constitución Política de 1917*. México DF: Biblioteca del Instituto Nacional de Estudios Históricos de la Revolución Mexicana. Reprinted 1984, PRI, Comisión Nacional Editora del CEN.

Sanders, Sol. 1986. *Mexico, Chaos at our Doorstep*. Lanham, MD: Madison Books.

Sanderson, Steven. 1981. *Agrarian Populism and the Mexican State*. Berkeley, CA.: The University of California Press.

Schteingart, Martha. 1978. *El Sector Capitalista Inmobiliario y las Formas de Apropiacion del Suelo Urbano: El Caso de Mexico*. Mexico, DF: El Colegio de Mexico (Centro de Estudios Economicos y Demograficos).

Schteingart, Martha. 1981. "Crecimiento Urbano y Tenencia de la Tierra: El Caso de la Ciudad Mexico." SIAP, *Revista Interamericana de Planificacion*. Vol. XV-60. (December, 1981): 168-181.

Schteingart, Martha. 1981. "El Proceso de Formacion y Consolidacion de un Asentamiento Popular en Mexico: El Caso de Netzahualcóyotl." *Revista Interamericana de Planificacion*-Vol. XV-57. (March, 1981): 100-114.

Schteingart, Martha. 1983. "La Promoción Inmobiliaria en el Area Metropolitana de la Ciudad de México (1960-1980)." *Revista Demografía y Economía* XVII: (January, 1983): 83-105.

Schteingart, Martha. 1983. *La Incorporacion de la Tierra Rural de Propiedad Social a la Logica Capitalista de Desarrollo Urbano: El Caso de Mexico*. Morelia, Mexico.: XIV Congreso Interamericano de Planificacion, SIAP: La Relación Campo-Ciudad 281-300.

Schteingart, Martha. 1988. "Mexico City in The Metropolis Era." *The Metropolis Era: Mega-Cities*. Ed. Matteei Dogan and John D. Kasarda. Vol.2. Newbury Park: Sage Publications: 268-293.

Scott, Ian. 1983. *Urban and Spatial Development in Mexico*. Baltimore, MD: The Johns Hopkins Press.

Senior, Clarence. 1958. *Land Reform and Democracy*. Gainsville, FL: University of Florida Press.

Silva Herzog, Jesus. 1964. *El Agrarismo Mexicano y la Reforma Agraria*. México, DF: Fondo de Cultura Economica, 2a Edicion.

Simpson, N. Eyler. 1952. "El *Ejido*: Unica Salida para México". *Problemas Agrícolas e Industriales de México*. Vol. IV, no. 4. Oct-Dec.

Simpson, N. Eyler. 1937. *The Ejido: Mexico's Way Out*. Chapell Hill, N.C.

SMP, Sociedad Mexicana de Planificacion. 1982. *Memoria de la Reunion Nacional sobre la Problematica del Suelo en Mexico*. Tlaxcala, Mexico.

Smith, M. David. 1979. *Where the Grass is Greener*. Baltimore, MD: The Johns Hopkins Press.

Solís, Leopoldo. 1970. *La Realidad Económica Mexicana: Retrovisión y Perspectivas*. México, DF: Siglo XXI.

Stavenhagen, Rodolfo. 1986. "Collective Agriculture and Capitalism in Mexico" in Hamilton, Nora and Timothy F. Harding. Edit. *Modern Mexico: State, Economy, and Social Conflict*. Beverly Hills: Sage Publications.

Stephens Rioja, Kathryn Irene. 1983. *Land Policy. Irregular Settlements and Metropolitan Growth in Newly Industrializing Countries. The Case of México City*. (Mimeo, Colegio de Mexico, ref. 301.36-58356L).

Stern, Claudio. 1983-84. "Population Growth, Migration and Rural-Urban Problems in Developing Countries." *World Futures*. Vol. 19- 3-4: 317-329.

Strong, Ann L. 1979. *Land Banking: European Reality, American Prospect*. Baltimore, MD: The Johns Hopkins Press, 1979.

Sudra, T.L. 1976. *Low-Income Housing System in Mexico City*. Cambridge, MA. Unpublished Thesis, MIT.

Tannenbaum, Frank. 1929. *The Mexican Agrarian Revolution*. New York: Macmillan.

Tannenbaum, Frank. 1933. *Peace by Revolution: An Interpretation of Mexico*. New York: Columbia University Press.

Tannenbaum, Frank. 1950. *Mexico the Struggle for Peace and Bread*. New York: Knopf.

Tannenbaum, Frank. 1966. *Peace by Revolution: Mexico after 1910*. New York: Columbia University Press.

Thurow, Lester C. 1980. *The Zero-Sum Society*. New York: Basic Books Inc.

Timberlake, Michael, ed. 1985. *Urbanization in the World Economy*. Orlando: Academic Press.

Tolley, George S. 1979. *Urban Growth Policy in a Market Economy*. New York: Academic Press.

Tolley, George S. and Vinod, Thomas. 1987. *The Economics of Urbanization and Urban Policies in Developing Countries*. A World Bank Symposium, Washington D.C: World Bank.

Turner, Alan. 1980. *The Cities of the Poor: Settlement Planning in Developing Countries*. London: Croom Helm.

Turner, John F.C. 1963. "Dwelling Resources in Latin America." *Architectural Design*, (August, 1963).

Turner, John F.C. 1971 *Government Policy and Lower Income Housing Systems in Mexico City*. Mexico, DF: Report to AURIS.

Turner, John F.C. 1975. "Asentamientos Urbanos no Regulados." *Desarrollo Urbano y Regional en América Latina*, eds. Unikel, Luís y Necochea Andrés. Mexico: Fondo de Cultura Económica.

Turner, John F.C. 1976-A. *Housing by People: Towards Autonomy in Building Environments*. London: Marion Boyars.

Turner, John F.C. 1976-B. *Uncontrolled Settlements, Problems and Policies*. Pittsburgh, PA: UN Seminar on Urbanization.

Turner, John F.C. and Fichter, Robert, eds. 1972. *Freedom to Build*. New York: Free Press, Collier Macmillan.

Tutino, John. 1986. *From Insurrection to Revolution*. Princeton NJ: Princeton University Press.

Unikel Luis. 1971 *Disyuntivas Sociales*. ed. Miguel S. Wionczek. Mexico: SEP/setentas No. 5.

Varley, Ann. 1985. "Urbanization and Agrarian Law: the Case of Mexico City." *Bull. Latin American Res.*, Vol.4- 1, (1985): 1-16.

Varley, Ann. 1985. " La Zona Urbano Ejidal y la Urbanización de la Cd. de México." Revista A15, Universidad Autonoma Metropolitana-Azcapotzalco, México.

Vining, Daniel R. 1986. "Population Redistribution Towards Core Areas of Less Developed Countries, 1950-1980." *International Regional Science Review*. Vol.10-1. (April 1986).

Vining, Daniel R. 1985. "The Growth of Core Regions in the Third World."

Scientific American. Vol. 252 - 4. (April, 1985).

Violich, Francis. 1981. "Mexico City and Mexico: Two Cultural Worlds in Perspective." *Third World Planning Review*. (Nov, 1981): 361-386.

Ward, Peter M. 1976. "The Squatter Settlement as Slum or Housing Solution: Evidence from Mexico City." *Land Economics*. 52. (August, 1976): 330-346

Ward, Peter M. 1981. "Mexico City." *Problems and Planning in the Third World Cities*, ed. Michael Pacione. New York: St. Martins Press: 29-63.

Ward, Peter M. 1983. "Land for Housing the Poor: How Can Planners Contribute?" in *Land for Housing the Poor*. ed. Angel, Shlomo et al. Singapore: Select Books.

Ward, Peter M. 1986. "The Politics of Planning in Mexico." *Third World Planning in Review*. (Aug. 1986): 219-235.

Ward, Peter M. 1990. *Mexico City. The Production and Reproduction of an Urban Environment*. London: Belhaven Press.

Whetten, L. Nathan. 1948. *Rural Mexico*. Chicago: The University of Chicago Press.

Williamson, Jeffrey G. 1986. *Migration and Urbanization in the Third World*. Discussion Paper, No. 1245. Cambridge, MA: Harvard Institute of Economic Research.

Womack, J. Jr. 1974. *Zapata en la Revolución Mexicana*. México DF: Siglo XXI.

Zaragoza, J.L., and Macías, R. *El Desarrollo Agrario de México y su Marco Juridico*. México, DF.